Radar and Electronic Navigation

G. J. SONNENBERG

BUTTERWORTHS

LONDON - BOSTON
Sydney - Wellington - Durban - Toronto

The Butterworth Group

United Kingdom	**Butterworth & Co (Publishers) Ltd** London: 88 Kingsway, WC2B 6AB
Australia	**Butterworths Pty Ltd** Sydney: 586 Pacific Highway, Chatswood, NSW 2067 Also at Melbourne, Brisbane, Adelaide and Perth
Canada	**Butterworth & Co (Canada) Ltd** Toronto: 2265 Midland Avenue, Scarborough, Ontario, M1P 4S1
New Zealand	**Butterworths of New Zealand Ltd** Wellington: T & W Young Building, 77–85 Customhouse Quay, 1, CPO Box 472
South Africa	**Butterworth & Co (South Africa) (Pty) Ltd** Durban: 152–154 Gale Street
USA	**Butterworth (Publishers) Inc** Boston: 10 Tower Office Park, Woburn, Mass. 01801

First published by George Newnes Ltd, 1951
Second edition 1955
Third edition 1963
Fourth edition by Butterworths, 1970
Fifth edition 1978
    Reprinted 1979, 1980

© G.J. Sonnenberg, 1978

> **British Library Cataloguing in Publication Data**
>
> Sonnenberg, Gerrit Jacobus
>   Radar and electronic navigation. — 5th ed
>   1. Electronic navigation
>   I. Title
>   623.89'3   VK560   77–30476
>
> ISBN 0–408–00272–7

Typeset by Scribe Design, Gillingham, Kent
Printed in England by Fakenham Press Ltd,
Fakenham, Norfolk

Foreword

The beginning of the last quarter of the twentieth century will enter the history of navigation as the period during which sea transport gradually developed into an integrated and industrialised system of transport.

It is obvious that the sea-going part of the transport chain from origin to destination must be a reliable and regular link. The achievement of reliability and regularity depends on crew capability, technical standards and navigational standards.

Electronic aids to navigation on board and ashore, together with a well educated and trained crew, should lead to the optimum efficiency of transport made possible by today's advanced and reliable nautical skills and knowledge. Therefore it is necessary that during the education of the crew considerable emphasis is put on the study of these aids. So it is not astonishing that now, within a quarter of a century, a fifth edition of *Radar and Electronic Navigation* makes its appearance.

Adapted to cover the latest stage of development in the electronic navigation field, this book will again find a place in nautical instruction. However, it should also be included in the standing inventory of every merchant ship. It is, in essence, like the Admiralty Pilot books: a guide showing the way through the fairway of electronic aids to navigation, changing course with time. As such this book is well worth consultation and study.

The Netherlands Maritime Institute wishes author and book the success they deserve.

Prof. Ir. W. Langeraar
General Manager, NMI

A. Wepster
Head of the Navigation
Research Centre, NMI

Preface

The continuing development of radar and other electronic aids to navigation has made it essential for me to rewrite this book almost completely. Its extent has been increased by the addition of chapters on the Omega system, the satellite system and integrated navigation systems. On the other hand the chapter on v.h.f. radiotelephony has been omitted, since this is a means of communication rather than navigation. As with previous editions, the book is intended to be used both for instructional purposes and in practice aboard ship.

The principal alterations in this new edition are as follows.

Consol. Since the importance of this system has decreased, the space devoted to it has been much reduced.

Decca. Although the Mark 12 receiver is still used on some ships, the description of it has been replaced by full coverage of the Mark 21 receiver.

Loran. More attention has been given to the Loran-C system.

Satellite system. Although this system is at present not much used by the Mercantile Marine, it is likely to play an important part in the future, especially if the plans for launching new navigation satellites are realised.

Radar. In connection with possible future applications, radar beacons have been more fully dealt with.

Integrated navigation systems. Such systems, based on computers, can increase the safety of navigation, and an account of these must therefore be given. Radars incorporating electronic plotting systems are also described.

Many people, to whom I am highly indebted, were kind enough to read and comment upon certain sections of the text, and to provide technical data. I would especially like to mention Mr Claud Powell of Decca Navigator Co. (who made a particularly important contribution); Mr P.J. Schreij of Observator Ltd, Rotterdam; Mr P.G. Sluiter of Shell International Oil Co., The Hague; and Mr A. Wepster, Head of the Navigation Research Centre of the Netherlands Maritime Institute, Rotterdam. As in the previous edition, Mr J. van Dijk, attached to the Nautical College at Rotterdam, has dealt with the subject of plotting.

Various organisations have put data, photographs and diagrams at my disposal. I can only mention Decca Navigator Co., The Hydrographer of the Navy, IBM (Navigation Department), Magnavox, the Netherlands Department of Waterways (North Sea Division), Observator Ltd, Philips, Plath of Hamburg, Siemens, Sperry Marine Systems, the US Coast Guard, and the US Naval Hydrographic Office.

Finally, I am extremely grateful to Mr A.W.V. Hoepermans, who collaborated with me in translating the book, and to Butterworths for their efforts in its preparation and production.

G.J.S.

Contents

1

Introduction

The growth of electronic navigation systems during the last thirty years
has been dramatic. Few scientific or industrial developments have ever
before expanded so rapidly, or found vital application so quickly and on
such a large scale. These electronic aids to navigation, which have already
proved of inestimable value in peace as in war, are still advancing in scope
and in reliability.

Radar, developed primarily as an instrument for detecting and ranging
in warfare, is of course the most important electronic aid to navigation.
Basically, radar employs very short electromagnetic waves and utilizes the
principle that these waves can be beamed, that they travel at a constant
speed in a straight line, and that they will be reflected by anything they
may meet. These waves are transmitted from the ship, and the echoes
received provide information that is presented visually on the screen of a
cathode-ray tube.

Decca, Loran and Omega use radio signals transmitted by stations of
known position. Special receiving equipment enables the navigator to
measure the difference between the times of arrival of signals from two
stations, and thus to determine his position.

The most promising system for position fixing is a satellite system that
has world coverage. A new underwater navigational aid is the Doppler log.
Direction finders and echo sounders have been used on board for many
years.

This chapter covers a certain amount of fundamental theory that must
be understood before these systems and their applications are described in
detail. Four systems (radar, Loran, the direction finder and the echo
sounder) make use of the cathode-ray tube, so an outline of its principles
of operation may be helpful. Consol, Loran and Decca are hyperbolic
navigational systems (that is, the lines of position provided by them have
a hyperbolic character), thus it is necessary to understand the construction
and uses of hyperbolas.

Because satellite position-fixing and the Doppler log are based upon the
Doppler principle, this is described. The chapter also includes a description

of the manner in which radio waves are transmitted through the atmosphere, the properties of the ionosphere, and the way in which radio waves are reflected back to Earth. These reflections have a great influence on the range over which the various position-fixing systems will function. First of all, however, a clear idea of alternating currents and voltages, phase difference, radiation and receivers must be given.

ALTERNATING CURRENTS AND VOLTAGES

A complete series of changes occurring so that the conditions at the end are identical with those at the beginning constitutes a cycle. In the graph of an alternating current shown in *Figure 1.1*, from t_0 to t_1 the current has one direction and from t_1 to t_2 the opposite direction. From t_0 to t_2 is one *cycle*. I_{max} is the maximum value or *amplitude*.

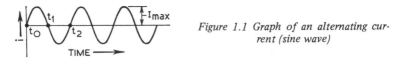

Figure 1.1 Graph of an alternating current (sine wave)

One cycle is taken to be equal to an angle, α, of $360°$. If we plot α on the horizontal scale instead of time, the form of the curve does not change. The instantaneous value of the current $i = I_{max} \sin \alpha$, hence the curve is called a *sine wave*. In what follows we will assume that alternating currents satisfy the formula above. A similar formula applies for an alternating voltage.

The number of cycles per second is the *frequency*. For example, the electricity supply in the United Kingdom is an alternating voltage with a frequency of 50 cycles per second or 50 hertz (50 Hz). For our purposes, the important frequency ranges for alternating voltages are 'audio frequencies' (about 30 to 18 000 Hz), which, when supplied to a loudspeaker or telephone, cause sound waves of the same frequency detectable by the human ear; and 'radio frequencies' (above 10 000 Hz), which are induced in radio-receiver circuits by the radio waves picked up by the aerial. Frequencies between 10 000 and 18 000 are (depending upon their origin and application) radio *or* audio frequencies.

One thousand hertz is conveniently written as one kilohertz (1 kHz). One million hertz is called one megahertz (1 MHz). In radar, frequencies of 10 000 MHz are common; to avoid the use of large numbers the term gigahertz may be used to signify 1000 MHz, so 10 000 MHz = 10 GHz.

Phase difference

Two alternating currents or voltages of the same frequency are *in phase* if

they reach their positive (or negative) maximum value at the same instant
of time and thus have the same direction at any moment.

Figure 1.2 shows two equal alternating currents; the current i_2 lags
behind i_1 by a quarter of a cycle or 90°. Should the phase difference be

Figure 1.2 Alternating currents of
same amplitude but 90° out of
phase

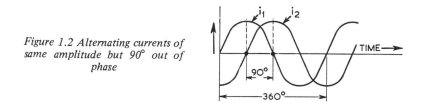

180°, the maximum positive swing of i_1 will occur at the same instant as
the maximum negative swing of i_2. Two alternating currents or voltages
can differ in frequency, amplitude and phase. *Figure 1.3(a)* shows two
alternating voltages of the same frequency but of differing phase and

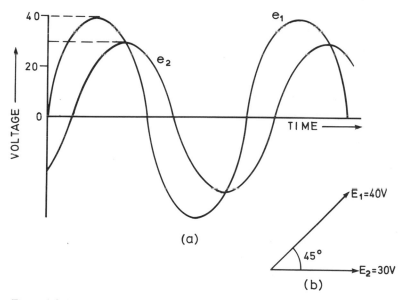

*Figure 1.3 (a) Alternating voltage e_1 (maximum value 40 V) leads e_2 (maximum
value 30 V) by 45°; (b) vector diagram of the two voltages*

amplitude. The voltage e_1 leads e_2 by 45° and has a greater amplitude:
40 volts compared with 30 volts.

Like forces and speeds, alternating currents and voltages can be indicated by *vectors*. If a vector represents, for instance, a force, the length of
the vector is proportional to the strength of the force and its direction

indicates the direction of the force. The length of a vector indicating an alternating voltage is proportional to the amplitude. In *Figure 1.3(b)* the two voltages of *(a)* have been plotted. The angle of 45° is their phase difference. One of the vectors, for instance E_2, may be given any direction; E_1 must then be plotted in a direction leading E_2 by 45°.

If there are two alternating currents or voltages of equal frequency in the same circuit, the resulting currents or voltages can be found as shown in *Figure 1.4*. The vectors I_1 and I_2 indicate two alternating currents; their resultant is the diagonal I_r of the parallelogram. Conversely, an alternating current or voltage can be resolved into two or more components. In *Figure 1.5* I_r is resolved into I_c and I_{ab}, and I_{ab} in its turn into I_a and I_b. Hence I_r is the resultant of I_a, I_b and I_c.

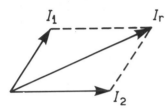

Figure 1.4 I_r is the resultant of I_1 and I_2; since I_1 and I_2 are sine waves I_r is also a sine wave

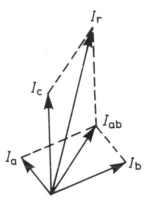

Figure 1.5 I_r is the resultant of I_a, I_b and I_c; conversely I_r can be resolved into, for instance, I_a, I_b and I_c

Due to the ease with which we can determine the resultant of two or more alternating currents or voltages, and also to the clarity with which the phase differences and the values of the currents and voltages are displayed, vector-diagrams (instead of graphs like *Figure 1.3(a)*) are generally used in the study of radio.

ELECTROMAGNETIC RADIATION FROM A RADIO-TRANSMITTER

According to the laws of electricity, a current passing through a wire produces a magnetic field around the wire. The number of magnetic lines of force (the strength of the field) is proportional to the current; see *Figure 1.6*. If we look in the direction of the current, the circular magnetic lines of force go in a clockwise direction. So if the current alternates, i.e. continually changes its magnitude (and periodically its direction), the magnetic field does the same. This field is called the *induction field*.

As proved theoretically by Maxwell in 1888, a current generates a *radiation field* simultaneously with the induction field. The radiation consists of magnetic lines of force of the same circular shape as those of the

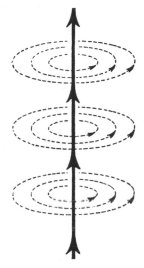

Figure 1.6 Induction field: magnetic lines of force produced by an electric current. If the current decreases, the radii of the circles decrease also

induction field. If the current decreases and ultimately disappears the magnetic lines of force of the induction field shrink, i.e. the radii of their circles decrease to zero; the radii of the lines of force of the radiation field, however, continue to increase if the current in the wire decreases to zero.

In *Figure 1.7* a radio-frequency current is supplied to the vertical wire of an aerial, O. Due to the rapid and continually repeated increase, decrease and reversal of direction of the current in the wire, a radiation field arises. The lines of force with an arrow in a clockwise direction were caused by a current that was directed away from us, and those with an opposite arrow by an opposite current. The lines of force between A and B, for example, were generated by one cycle of the alternating current; this distance is called the *wavelength*. In one second the lines of force travel a distance of 300 000 km. This propagation speed is always the same irrespective of the frequency. The higher the frequency the stronger the field, however.

If the frequency is f Hz there are f cycles per second, and thus f waves per second are generated. These f waves travel a distance of 300 000 km in one second, so the length λ (lambda) of one wave is:

$$\lambda = \frac{300\,000}{f} \text{ km} = \frac{300\,000\,000}{f} \text{ m}$$

With this formula the wavelength can be calculated if the frequency is known, and vice versa. Normally the radiation field is indicated by its frequency.

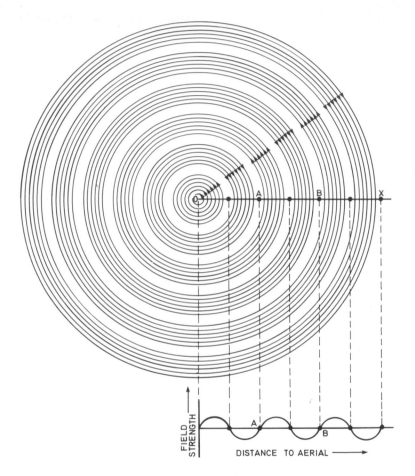

Figure 1.7 Radiation field: magnetic lines of force generated by a high-frequency current in a transmitting aerial, O. The radii of the circles increase at a speed of 300 000 km/s; AB is the wavelength. The graph shows the field strength along the line OX at a given moment

The wavelength is also the distance covered by a line of force or a wave during one cycle. If we indicate the speed of propagation by c, the formula above ($\lambda = c/f$) is also applicable to sound waves in the air and in the water. For sound in the air c is about 330 m/s and for sound in the water c is about 1500 m/s.

Besides magnetic lines of force there are always electrical lines of force in the radiation field of a transmitter. The latter have a direction *perpendicular* to the magnetic lines. The combination of the two fields is called the *electromagnetic radiation* field.

Unlike the radiation field, the induction field decreases very rapidly as distance from the aerial increases; it may be neglected at a distance of only one wavelength from the aerial. Hence the induction field plays no part in radio communication.

Polarization

The magnetic lines of force need not lie in a horizontal plane, nor the electrical lines in a vertical plane. If the electrical lines are in the vertical direction the field is said to be *vertically polarized,* if they are in the horizontal direction the field is *horizontally polarized.* As appears from *Figure 1.7*, the strength of the field is not the same everywhere at a certain point of time. Going from O (the transmitting aerial) to X, the field increases, decreases, changes direction, increases and decreases again, etc. The graph at the bottom of the figure, a sine wave, shows the field strength versus the distance from the transmitter. As the speed of the field is constant, the graph of the field strength *at a fixed point* is also a sine wave.

It goes without saying that the radiation field at a great distance from the transmitter will, as the result of attenuation, be very weak. If we make the alternating current in the aerial stronger, the field will become stronger too and the distance that the transmitter can cover (the range) will increase.

Frequency bands

A radiation field is characterized, at least for our purposes, mainly by its frequency; this frequency is the same, of course, as that of the aerial current. The complete frequency spectrum usable for radio lies between about 300 GHz and 10 kHz. On an international basis it is allocated in frequency-bands and assigned for numerous purposes.

For present radio-navigation systems the following frequency bands are used:

Omega system:	approx. 10 to 14 kHz
Decca system:	approx. 65 to 135 kHz
Loran system:	100 kHz or approx. 1900 kHz
Radio beacons:	approx. 300 kHz
Consol system:	approx. 300 kHz
Satellite systems:	approx. 150 MHz and approx. 400 MHz
Radar:	approx. 10 GHz and approx. 3 GHz

Röntgen, ultra-violet, light, infrared and heat radiation are also electro-magnetic fields. They are basically the same as the radiation field of an aerial; they differ in frequency, however, and consequently in their propagation.

Radio reception

According to the principles of electricity a voltage is generated in a conductor if the latter is moved in such a way that it intersects magnetic lines of force. The direction of the voltage is perpendicular to the direction of the lines of force and also to the direction of movement of the conductor.

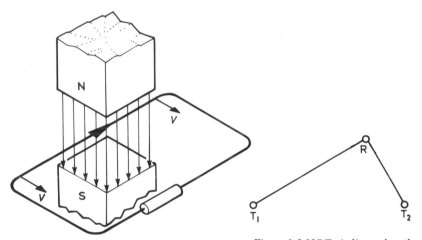

Figure 1.8 Voltage induced in wire by magnetic field is proportional to field strength

Figure 1.9 If RT_1 is ¾ wavelength longer than RT_2, the two voltages induced in receiving aerial R by transmitting aerials T_1 and T_2 will have a phase difference of ¾ × 360° = 270°

Figure 1.8 shows a wire that is moved to the right, at a velocity v, in the magnetic field between the North and South poles. The wire, the field and the direction of movement are perpendicular to each other. The generated voltage is directed backwards.

According to the same principle, an alternating voltage will be generated in a vertical wire located within the range of a transmitting station. The wire functions as a *receiving aerial*. The frequency of the voltage is, of course, the same as that of the field, and hence of the transmitting-aerial current. The alternating voltage is very weak and needs to be amplified considerably in the receiver to which the receiving aerial is connected. For radio communication this voltage must be converted to a form that is perceptible to one of our sense organs, e.g. into a sound or an image on a screen. This will be described later on.

In connection with applications we can now point to the following:

(a) If we suppose that in *Figure 1.7* receiving aerials are located at A and B, the generated alternating voltages will have a phase difference of exactly 360° (or 0°, which is the same thing), so they are in phase. The same holds good for any two points that are one wavelength apart.

(b) If one aerial is at A and the second at a distance of a quarter of a wavelength to the right of A, the difference in phase between them is ¼ × 360° = 90°.

(c) As regards the phase of the voltage, it does not make any difference whether a receiving aerial is at A or somewhere else *where the distance from the transmitting station is the same*.

(d) In *Figure 1.9* T_1 and T_2 are transmitting stations, transmitting on the same frequency. Moreover, their aerial currents are in phase. If the distance of the receiver R to T_1 is ¾ wavelength longer than to T_2 the two voltages generated in R will have a phase difference of ¾ × 360° = 270°. The voltage caused by T_2 leads the other voltage.

If there are two voltages of the same frequency, as in case (d), the resultant voltage can be found by vector diagram (*Figures 1.4* and *1.5*). Normally, however, we consider the two voltages separately.

Noise

In conducting material, e.g. a metal wire, some of the electrons have escaped their orbits around the nuclei and move freely inside the wire. Due to collisions with other particles their direction of movement changes. Normally these 'free electrons' do not leave their conductor, however, so there is no 'emission'.

The number of electrons moving during a very short period towards one end of the wire is not equal to the number moving towards the other end during the same period; the result is an electric current, although extremely weak. During the next short period, however, this current might have the reverse direction. As a result there arise between the ends of the wire electric voltages of a very irregular character. These voltages increase with temperature. If a radio receiver has sufficient amplification, the voltages become audible as background noise; hence the name *noise*.

Another cause of radio interference is radio waves, sometimes very strong, caused by discharges in the atmosphere. Some (mostly tropical) regions are notorious for this interference, called *atmospherics*. Atmospherics can be reflected by the ionosphere (see later) and hence can impede reception at very long distances.

Precipitation particles (rain, hail) may in special circumstances become electrically charged. If they make contact with the receiving aerial they discharge to earth via the aerial and receiver. This so-called *precipitation static* can cause serious interference.

A further cause of interference is sparking switches and motors, especially high-frequency apparatus, which induce voltages in the receiver and in the receiving aerial ('man-made' noise). This interference is only inconvenient up to a short distance from its source. Its effects can therefore be avoided by mounting the receiving aerial in a high position above the 'interference level'. The aerial wire below this level, its connection to the receiver and the whole receiver itself should then be surrounded by conductors (shielding cables, metal cases etc.) directly connected to earth.

A characteristic of noise, irrespective of its cause, is that it appears on all frequencies. Hence by making the receiver highly selective, or in other words by 'opening the door' of the receiver only for the frequency of the desired station, much noise can be got rid of.

If we receive, besides noise, a transmission of sufficient power, the noise disappears. This is due to the fact that the automatic volume control of the receiver automatically decreases the degree of amplification as the strength of the signal increases. As the noise is normally weaker than the signal, the noise is insufficiently amplified to become audible. If, however, the transmitter signals are only as strong as (or even weaker than) the noise, the signals cannot be separated from the noise.

The greater the distance from the transmitting station the weaker the signal, of course. In principle noise is equally strong anywhere on earth, however; hence at a certain distance from the transmitter the signal is as strong as the noise. Greater amplification by the receiver will be of no avail then. Only an increase of the power of the transmitter could improve the reception.

Oscillators

Both in transmitters and in receivers oscillators are needed to generate electrical oscillations. Some types generate a frequency that is adjustable; with other types this is not possible. An example of the latter type is the quartz oscillator; this contains a quartz plate (about the size of a coin) that determines the frequency.

A very special oscillator is the atomic oscillator, also called atomic clock. The frequency it generates is determined by atoms of rubidium or caesium. Its relative accuracy, which is the frequency-deviation $\Delta f/f$, amounts to only about 10^{-11}. This means that after 10^{11} oscillations the oscillator has generated about one oscillation more or less than it should have done. Compared with all other oscillators this accuracy is exceptionally good. The use of atomic oscillators will make possible in the future radically new radio position-fixing systems.

Flywheel oscillators

In the receivers of some systems (e.g. Decca, Loran) there are special oscillators that generate oscillations with not only the same frequency as those received from an outside source but also the same phase; the oscillations are *phase-locked* to those of the source. This would not be possible if the oscillator generated its oscillations independently (just as the tick of a nonelectrical clock will not coincide with that of another clock for a long time).

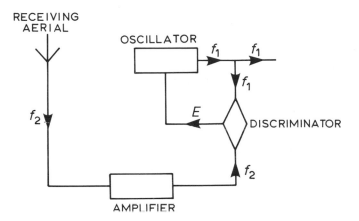

Figure 1.10 Flywheel oscillator: a phase difference between f_1 and f_2, both supplied to the discriminator, gives rise to a direct voltage E, which forces the oscillator to change the frequency and phase of f_1 to that of f_2

The block diagram *Figure 1.10* shows the method used. The oscillations f_2 received by the aerial are amplified and supplied to a *discriminator*, which also receives the oscillations f_1 from the oscillator. If the phase-difference between f_1 and f_2 is not zero a *direct voltage*, called error voltage E, is produced in the discriminator. The magnitude of E is determined by the difference in phase between oscillations f_1 and f_2. The voltage E is supplied to the oscillator, where it increases or decreases the frequency (i.e. the speed of generating oscillations f_1) in such a way that the difference in phase becomes zero. When this has been achieved the error voltage becomes zero too.

A flywheel, left to itself, will keep the number of its revolutions per minute constant for a short time. In the same way our oscillator, called a *phase-synchronous, flywheel* or *phase-locked* oscillator, will not change its phase for a short time if there is a short break in the signal f_2. In other words, the oscillator retains the phase of the outside source, for instance a transmitter, during the short intervals between its signals. So it is not necessary for the oscillations of the transmitter to arrive continuously.

The received frequency may be 'contaminated' by noise and interference from some other origin. Frequency f_2 with its contamination cannot get to the oscillator, and hence the oscillations generated are free from interference. Noise, atmospherics etc. also have practically no influence on the error voltage that corrects the phase of the oscillator. This means that noise etc. will not influence the correct functioning of the oscillator. For the receivers of some position-fixing systems this is very important.

Detection

As mentioned before, only frequencies between about 30 Hz and 18 kHz are audible when converted to sound waves. With a few exceptions (e.g. the Omega system) the received frequencies are considerably higher, so they have to be converted into audible frequencies.

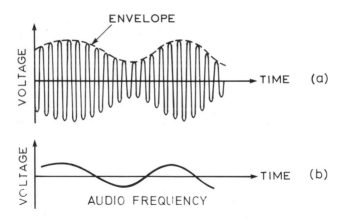

Figure 1.11 A detector circuit converts the modulated radio-frequency voltage of (a) into the audio frequency voltage of (b); the shape of (b) is that of the envelope of (a)

For this conversion the *detector stage* is incorporated in the receiver, comprising among other things a diode. The voltage before detection is of high frequency with varying amplitude, as in *Figure 1.11(a)*; the voltage after detection is of low frequency, as in *Figure 1.11(b)*, with the shape of the envelope of *Figure 1.11(a)*.

Beats

Another process to convert higher into lower frequencies is as follows. In *Figure 1.12(a)* a frequency f_1 is shown and in *(b)* a frequency f_2; the latter might, for instance, be an auxiliary frequency, generated in an oscillator. The difference between the frequencies is small compared to the frequencies themselves. The two frequencies are brought together in the same circuit; their resultant can be found by adding, for any point of time, the instantaneous values of both, taking into account the direction of the voltages. So AB + CD = EF and GH − IJ = KL. The results are *beats*, shown in *Figure 1.12(c)*; their amplitudes increase and decrease regularly. By a beat we understand the oscillation from t_1 to t_2. The numbers of beats per second is called the *beat frequency*; it can be proved that this is exactly equal to the difference of the two frequencies f_1 and f_2.

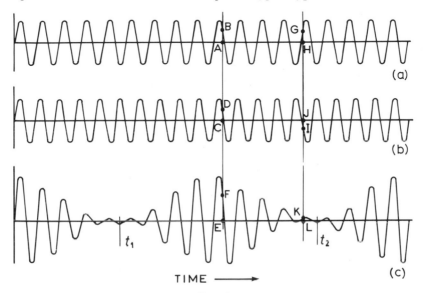

Figure 1.12 (a) Frequency f_1; (b) auxiliary frequency f_2, generated in an oscillator; (c) resulting oscillation, AB + CD = EF and GH − IJ = KL. From t_1 to t_2 is one beat, and the beat frequency equals the difference between f_1 and f_2

The resulting frequency can be detected as explained before to obtain the envelope frequency, which is, of course, much lower than f_1 or f_2. So if f_1 = 500 kHz and f_2 = 501 kHz the envelope or beat frequency will be 501 − 500 = 1 kHz. It is clear that a variation of f_2 will bring about the same variation in the beat frequency. If, for instance, f_2 in our example was not 501 but 503 the beat frequency would be 503 − 500 = 3 kHz instead of 1 kHz. To obtain a sound frequency of, say, 5 kHz the oscillator would have to produce a frequency of 505 kHz.

The electronic process of obtaining a frequency that is exactly equal to the difference between two other frequencies is used, for instance, in broadcast receivers to convert a received high frequency to an audible frequency. It is also used in special receivers for position-fixing purposes.

Transmitters

The function of a radio transmitter is to produce radio-frequency currents; normally the frequency and the strength of the current can be varied between two limits. The currents are supplied to the aerial, which converts them to electromagnetic waves.

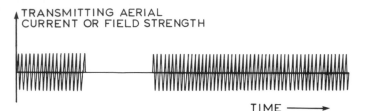

Figure 1.13 *Aerial current or field strength of a continuous-wave transmitter if the Morse letter A is transmitted*

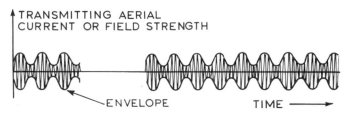

Figure 1.14 *Aerial current or field strength of a modulated-continuous-wave (or modulated-telegraphy) transmitter if the Morse letter A is transmitted*

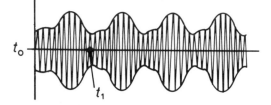

Figure 1.15 *Aerial current or field strength of a telephony-transmitter; the radio-frequency amplitude is modulated by the audio-frequency oscillations generated in the micro-phone*

We can divide transmitters into four types:

(a) Transmitters that generate *continuous waves* (c.w.). *Figure 1.13* is a graph of the current (and also of the radiated field) if the Morse letter A (dot dash) is transmitted.

(b) Transmitters generating *modulated continuous waves* (m.c.w.). The amplitudes of the high-frequency cycles are varied or, as it is called, *modulated*; the envelope in *Figure 1.14* is a sine wave. Again the letter A is being transmitted.

(c) Transmitters for *radio-telephony* (see *Figure 1.15*).˙The sound, for instance the human voice, is converted in the microphone into low-frequency electrical oscillations, which modulate the high-frequency oscillations. A radio-telephony transmitter does not transmit on one frequency only but occupies a frequency-band of about 9 kHz. The frequency in the middle of this band is by far the strongest and is called the *carrier*.

(*d*) Transmitters generating *pulses* (e.g. radar and Loran transmitters). Every pulse is very short compared with the intervals between the pulses (see for instance *Figure 9.2(c)*). Because of this, the frequency-band occupied by this transmitter is much wider than that of a radio-telephony transmitter.

Simple receiver

Figure 1.16 is a block diagram of a very simple three-stage receiver. The frequency received by the aerial is extremely weak, and is amplified in the first stage of the receiver and detected in the second stage. Detection changes it from a radio frequency to an audio frequency. In the third stage it is again amplified, and finally it is supplied to the loudspeaker or telephone, which converts it into sound.

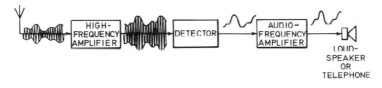

Figure 1.16 Simple three-stage receiver; the detector converts radio frequencies to audio frequencies

The envelope of the received frequency is in fact the graph of the sound produced in front of the microphone at the transmitter. The same graph is recovered in the receiver, and therefore the same sound as picked up by the microphone will be reproduced by the telephone or loudspeaker. In *Figure 1.16* the receiver has only one radio-frequency and one audio-frequency amplification stage; as a rule there are more stages, however.

Continuous-wave signals cannot be made audible by this receiver, because the amplitude of the received signal does not vary, except at the end and commencement of the dots and dashes. To make c.w. signals audible the receiver must be equipped with a radio-frequency oscillator (*Figure 1.17*). The frequency (f_2) of this oscillator differs somewhat from the signal frequency f_1. Frequencies f_1 and f_2 are supplied to the same circuit (called *mixing*) and detected. The result is a signal with a frequency equal to $f_1 - f_2$, which is supplied (after audio-frequency amplification) to the telephone or loudspeaker. It is clear, therefore, that a sound with a frequency equal to the beat frequency will be heard.

If, for instance, the signal frequency is 300 kHz (radio beacons) and the receiver-oscillator frequency f_2 is 301 kHz, the sound-frequency will be

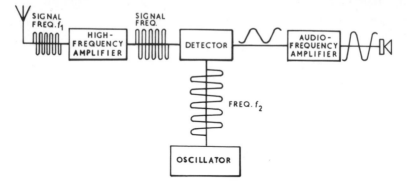

Figure 1.17 Reception of continuous waves: the auxiliary frequency f_2 is mixed with the signal frequency f_1, and the beat frequency $f_1 - f_2$ is detected; this gives the sound frequency, which is adjustable by varying f_2

301 − 300 = 1 kHz. Varying the oscillator frequency from, say, 490 to 510 kHz decreases the sound frequency from 10 kHz to zero and then increases it again to 10 kHz.

Superheterodyne receiver

The receivers so far described, called direct receivers, have several draw-backs and have long been superseded by the *superheterodyne receiver*. *Figure 1.18* is a simplified block diagram of such a receiver. The 'mixer

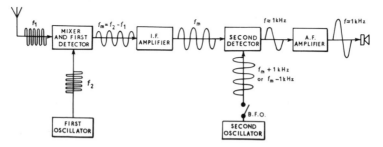

Figure 1.18 Superheterodyne receiver: in the first stage the received frequency f_1 is mixed with the first oscillator frequency f_2 and detected to give the intermediate frequency f_m; f_2 is adjusted so that f_m remains the same whatever the value of f_1. For reception of modulated signals the second oscillator should be disconnected by means of the b.f.o. switch

and first detector' is supplied with two frequencies, the received frequency f_1 and the frequency f_2 that is generated by the first oscillator. From these two a high frequency called the *intermediate frequency*, $f_m = f_2 - f_1$, is obtained. In the same way as in the direct receiver of *Figure 1.17*, f_m is then amplified, detected etc. A characteristic of the superheterodyne receiver is the fact that *the intermediate frequency is always the same, irrespective of the received frequency* f_1. This is achieved by adjusting f_2 to such a value that its difference from f_1 remains the same. If we want to receive, for instance, 500 kHz, and the intermediate frequency f_m is fixed at 125 kHz, f_2 must be equal to 500 + 125 = 625 kHz; if the frequency to be received is 8 MHz, f_2 must be 8.125 kHz.

The great advantage achieved is that the intermediate-frequency circuits can remain permanently tuned to the same frequency and *do not need variable capacitors* for tuning (see below). The circuits increase the selectivity, and cumbersome tuning becomes superfluous. There are other advantages that need not be discussed here.

If the received frequency is modulated, the intermediate frequency has exactly the same modulation and the second oscillator shown in *Figure 1.18* must be switched off. For reception of continuous waves, however, the intermediate frequency is not modulated and the second oscillator has to be switched on, using the 'beat frequency oscillator' (b.f.o.) switch. If, for instance, the b.f.o. frequency is 1000 Hz higher or lower than the intermediate frequency, the frequency of the sound produced is 1000 Hz. By changing the frequency of the second oscillator the pitch of the sound can therefore be varied.

Sensitivity and selectivity

By the *sensitivity* of a receiver we mean its capacity to make audible very weak signals, in other words to *amplify* the signals sufficiently. The amplification can be adjusted by the *manual volume control* (m.v.c.) and also sometimes by a *radio-frequency gain* (r.f. gain) control.

A receiver must also be able to select the desired station from those transmitting on neighbouring frequencies. This property of a receiver is called *selectivity*; it can be obtained by making use of the *electrical resonance phenomenon* in tuned circuits. Such circuits discriminate between the signals of different stations on the basis of their frequencies. By using several tuned circuits, each consisting of a capacitor and a coil, selectivity can be increased. The capacitor is usually a variable one and has to be tuned to the desired station.

By means of the 'channel switch' or 'band-change switch' the capacitor and/or the coil can be put to other settings or replaced by other capacitors or coils for the purpose of tuning to another frequency or another frequency band.

Some receivers have a switch to increase the selectivity. Often insufficient use is made of this possibility of decreasing interference. To take a bearing from a radio beacon, for instance, the selectivity switch of a certain type of direction finder should be put first in the position 'broad' or 'wide'. After the radio beacon has been identified the switch should be set to the position 'narrow' and, when the tuning control has been adjusted again, to the next position of the switch, 'filter'.

In order to avoid variations in the volume of the sound, which may be caused by fading (see page 40), receivers are equipped with *automatic volume control* or *automatic gain control*. By means of a.v.c. or a.g.c. the amplification is automatically *increased* as the arriving signal becomes *weaker* and *decreased* as the signal becomes *stronger.*

PROGRESS IN ELECTRONIC ENGINEERING

In 1930 a broadcast receiver contained about seventy components (resistors, valves, capacitors, coils etc.). A modern colour television contains about 7000 components and a large computer perhaps 200 000. This trend has certainly not finished, and holds good for electronic apparatus and systems aboard ships and aircraft.

If the components still had the same dimensions as, for instance, in 1930 it would be difficult or impossible to construct some of the present advanced electronic apparatus and systems. A minimising of the dimensions was needed. Apart from this special problem the reliability of electronic systems, especially on board ships, had to be increased. This can be demonstrated by the following example. Suppose that the broadcast receiver of 1930 with its seventy components malfunctioned once every two years, then the present colour television with its 7000 components (if constructed in the same way) would malfunction fifty times a year. Hence it is clear that apparatus and systems are needed with a considerably greater reliability than in the past.

It follows that there was a need for two developments: miniaturisation and greater reliability. The fruits of these developments are printed circuits, semiconductor devices, integrated circuits and microprocessors.

Printed circuits

Malfunction of electronic equipment is often caused by bad connections between the many components; this is especially the case with soldered connections. An improvement in this respect started in about 1950 when plug-in *printed circuit boards* (PCBs) were introduced. By a special process resembling printing, copper 'wiring' is laid down on one side of a board of insulation material (sometimes flexible); the components are mounted on the other side of the board, and are connected to the 'wiring' via holes in the board. This process provides a higher reliability than soldered connections. *Figure 1.19* shows such a PCB being taken out of the case; the wiring is connected to plugs (not visible in the figure) which automatically

Figure 1.19 Plug-in module being removed (courtesy Siemens)

make contact via the plugs of other PCBs to the wiring and components of those other PCBs. The manual for the apparatus or system describes how to ascertain which plug-in board must be replaced in the event of malfunctioning or breakdown.

Semiconductors

A very simple definition of a semiconductor is: a material that has a resistivity between that of an insulator and a good conductor. At present the term 'semiconductor' is restricted to silicon and germanium, which are 'doped' with a relatively very small number of atoms (only about 1 atom in 10^6 atoms of silicon) of other elements such as indium, arsenic and gallium. (The latter elements have some atomic resemblance to silicon.)

Nowadays silicon is generally used as the doped semiconductor material. Depending on the kind of added element and its quantity, the silicon becomes p- or n-type and its resistivity is altered consequently. The principles of semiconductors will not be explained here, but some of their applications in navigational equipment can be mentioned very briefly.

P-type silicon has a fairly high resistivity and is suitable for making *resistors* (see *Figure 1.20*). It must be emphasised that the vertical scale in

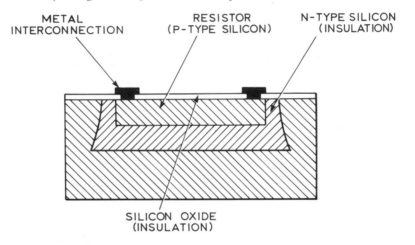

Figure 1.20 Diagrammatic section through semiconductor resistor

this figure has been modified in the interest of clarity. The resistor structure itself is, in reality, about 0.01 mm thick compared with about 0.2 mm for the whole so-called 'chip'.

A piece of silicon consisting partly of p-type and partly of n-type silicon presents very high resistance to a current in one direction, whereas its resistance to a current in the reverse direction is practically zero. Such a *semiconductor diode* serves, for instance, as a detector in receivers.

The *transistor*, the most important semiconductor device, has superseded the valve to a large extent and is used, for instance, as an amplifier.

There are also *semiconductor capacitors* and *semiconductor solar cells*. The latter are used in satellites for converting solar light directly into electrical energy.

A *thermistor* is a semiconductor resistor that is sensitive to temperature. The temperature can be brought about either by an electric current through the thermistor, or by the thermistor's surroundings. As an example of the latter case, thermistors are used in Doppler-log transducers:

the thermistor resistance is a function of water temperature and hence of the speed of sound through the water, the exact value of which must be known for the Doppler log.

Note that electronic apparatus built with semiconductor devices is known as *solid-state* equipment.

Integrated circuits

An *integrated circuit* (i.c.) is basically a small piece or 'chip' of silicon with several doped layers (of different shapes and extending to various depths) forming the components, and with a metallic interconnection pattern on the surface. An i.c. might be 0.2 mm thick and occupy an area of up to 20 mm^2 with hundreds of transistors and other semiconductor devices, all

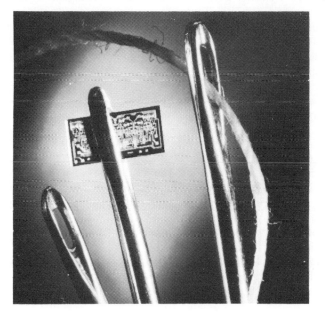

Figure 1.21 Integrated circuit (courtesy Mullard)

formed in the same operation and inseparably connected. By means of the present technology of *large-scale integration* (LSI) it is possible to construct integrated circuits in chips with even smaller dimensions and with thousands of components. See *Figure 1.21*.

For the following main reasons semiconductor integrated circuits are important:

(a) they are small and light, two major requirements in aerospace applications, for example;

(*b*) they are potentially more reliable than conventional circuits, as they do not rely so heavily on soldered or wire-wrapped connections;

(*c*) being made by mass-production techniques they are inexpensive;

(*d*) it is possible to produce special circuits that cannot be made by conventional techniques.

Microprocessors

A microprocessor is a single integrated circuit that performs all the arithmetic functions in a microcomputer, which is employed when the power and flexibility of a minicomputer is not required to its full extent and thus is not economically justified. Though a microprocessor is normally programmed for a special purpose it can still be reprogrammed, and hence does have some flexibility. There are many sectors where microprocessors will be used in future, for instance in telecommunications, test and measuring instruments, and electronic navigational aids. The computers of some satellite receivers for position-fixing have already been replaced by microcomputers.

Reliability, maintenance and repair on board

Electronic equipment on board is, more than that on shore, exposed to extreme temperatures and great temperature variation, moist air, salt, and mechanical shocks and vibrations. The possibilities of repair, should breakdown or malfunction occur, are limited both at sea and in many ports, among other reasons because spare parts are not always immediately available. It is therefore necessary that:

(*a*) electronic apparatus and systems for use on board should have great reliability;

(*b*) spare parts for essential components or complete units should be on board (some of them are officially obligatory);

(*c*) by means of the manual, personnel with only a restricted knowledge of electronics should be able to maintain and to adjust the apparatus, to localise errors and, if necessary, to perform repairs.

In modern apparatus valves are mostly replaced by transistors. This means an important increase in component lifetime and in reliability; moreover, the technical performance of transistors does not change over their lifetime. For a further increase in reliability, duplication of subunits, units and even complete systems is applied.

As a measurement of the reliability of a system the *mean time between failures* (MTBF) can be used. A better criterion, however, is the *down-time*, i.e. the time during which a system does not function during, say, 100 'working hours'. If, for instance, the system has a 30-minute breakdown during 2000 working hours, the down-time is $30/(2000 \times 60) = 1/4000 = 0.025$ per cent.

DOPPLER EFFECT

The Doppler log and the satellite position-fixing system are based on measurement of the Doppler effect, described now.

In 1842 in Vienna, Doppler pointed out that a star moving towards us must appear to be a different colour than when moving away from us. Buys Ballot proved experimentally in 1845 that an approaching source of sound is heard to have a higher frequency when its source is approaching the observer than when the source is stationary, thus proving the existence of the Doppler effect.

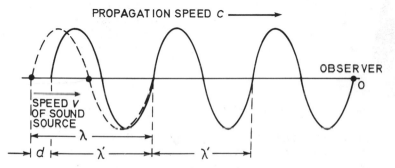

Figure 1.22 Doppler effect: as a result of the speed of the sound source, each wave is shortened from λ to λ' and the frequency received at O is increased

The effect can be calculated as follows. Suppose that in *Figure 1.22* the velocity of a sound source is v and the speed of propagation of the sound waves is c. Every wave is shortened because of the displacement of the source by d. This shortening is equal to the distance the source has moved during the time of generating one wave, i.e. during the time of one cycle T.

As $T = 1/f$, $d = vT = v/f$. Hence the wavelength λ' becomes $\lambda' = \lambda - (v/f)$, and the frequency heard is:

$$f' = \frac{c}{\lambda'} = \frac{c}{\lambda - (v/f)} = \frac{cf}{\lambda f - v} = f\frac{c}{c - v} \tag{1.1}$$

So the frequency f' is higher than f. The change in f is called the *Doppler shift*. If the source moves *away* from the observer, v becomes negative in equation 1.1 and:

$$f' = f\frac{c}{c + v}$$

The frequency f' is now lower than f.

Suppose now that the observer, at a velocity v, approaches a stationary source of sound. In that case the number of waves reaching the observer per second is equal to the number of waves generated by the source per

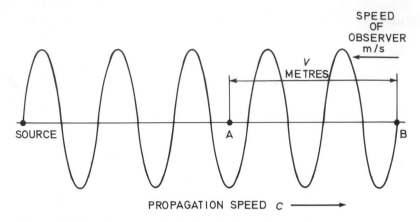

PROPAGATION SPEED c ⟶

Figure 1.23 An observer moving in the direction of a sound source receives, per second, the number of waves produced by the source per second plus the number of waves along the distance he travels per second

second, plus the number of waves on the line AB in *Figure 1.23*. AB is the distance of v metres covered by the observer per second. The number of waves at AB is v/λ. So the received frequency is:

$$f_r = f + \frac{v}{\lambda}$$

and since $1/\lambda = f/c$:

$$f + \frac{fv}{c} = f\left(1 + \frac{v}{c}\right) = f\frac{c+v}{c} \tag{1.2}$$

A special case occurs if the source and the observer are both moving towards a reflecting plane. We should then replace f in equation 1.2 by f' from equation 1.1, so that the received frequency f_r becomes:

$$f_r = f'\frac{c+v}{c} = f\frac{c}{c-v}\frac{c+v}{c} = f\frac{c+v}{c-v} \tag{1.3}$$

CATHODE-RAY TUBE

To clarify the principle of operation of the cathode-ray tube, it is necessary first to consider the simple diode valve.

If a metal wire (the filament) placed in a vacuum tube is heated by passing an electric current through it until it becomes incandescent, the liberated electrons in the metal reach such a high speed that many are detached from the filament and hurled into the vacuum space around it. In this way the wire is surrounded by a cloud of electrons. If a positive voltage is applied by means of a battery to a metal plate placed in the vicinity of the filament, the electrons will be attracted to this plate or

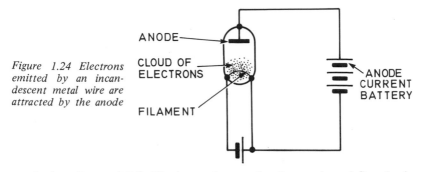

Figure 1.24 Electrons emitted by an incandescent metal wire are attracted by the anode

anode (see *Figure 1.24*). They are taken up by the anode and flow back through the anode-current battery to the filament.

A simple form of cathode-ray tube is shown in *Figure 1.25*. Again the principle of operation is based on the emission of electrons by a heated filament. In this case, however, the anode to which they are attracted has a hole in it through which a beam of electrons will pass at a very high speed

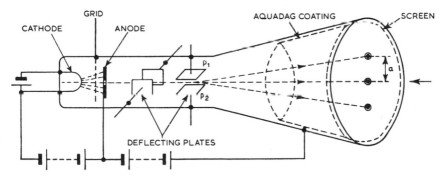

Figure 1.25 Cathode-ray tube. Some of the electrons pass through the hole in the anode to form an electron ray. These electrons bombard the screen, causing a luminous spot. A voltage between p_1 and p_2 moves the spot up or down

(the speed depending on the anode voltage), bombarding the screen at the end of the tube. The screen is coated with a material that glows under electron bombardment, producing a small, bright spot of light if the screen is observed from the direction of the arrow on the right-hand side of *Figure 1.25*. The deflecting plates enable the electron beam to be directed to various points on the screen to trace out patterns. As the glow of the screen coating does not cease instantly when the beam moves from a particular point, a complete picture can be built up on the tube screen (as in the case of television receivers). How long this afterglow continues depends on the type of screen coating used; it may be anything from a few microseconds to several minutes (so-called short- and long-persistence tubes).

The left-hand part of *Figure 1.25* where the electron ray originates is sometimes called the 'electron gun'.

The electrons bombard the screen with such an intensity (the speed can be thousands of kilometres per second) that other electrons are dislodged from the screen material; this is called 'secondary' emission. These secondary electrons are attracted to a conductive coat of graphite, the Aquadag

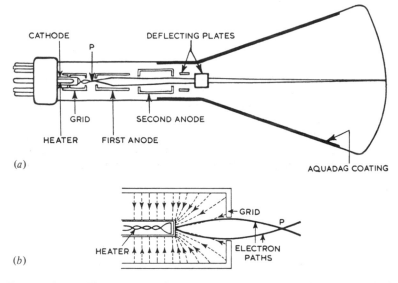

Figure 1.26 (a) Electrostatic cathode-ray tube; (b) heater, cathode and grid of a cathode-ray tube

coating (see *Figure 1.26(a)*), to which a high positive voltage is applied so that it will 'collect' the secondary electrons. On average, the screen loses in this way as many electrons as it receives from the cathode. It is also the task of the Aquadag coating to shield the interior of the tube from external electric and magnetic fields, preventing these fields from influencing the path of the electrons.

Because of the high vacuum inside the tube, the glass is subjected to very considerable pressure from the air outside. Therefore great care is needed in handling tubes and in working near them if not protected. If an 'implosion' occurs not only can the glass cause injuries, but in addition the material of the fluorescent screen can cause blood-poisoning in a wound.

Figure 1.26 shows a cathode-ray tube in more detail. The filament or heater shown at *(b)* is inside the cathode, which is a small box that emits electrons from its outer surface. The grid consists of a cylinder surrounding the cathode.

Electrons are emitted by the cathode at various speeds and are repelled by the negative grid (as it has a similar charge). The electric forces are indicated in *Figure 1.26(b)* by broken lines (the arrows show the direction of the electric field, and therefore have a direction opposite to the direction of the forces). The electrons are attracted by the positive anodes shown in *Figure 1.26(a)*. As a consequence, the electrons follow curved paths which all go through point P irrespective of their initial velocities. There are two (sometimes three) anodes of cylindrical shape. Unlike those in diode valves, these anodes do not pick up electrons.

Focusing

By applying to the grid, anodes and Aquadag coating the correct voltages with respect to the cathode, the electrons will follow the lines shown. After passing point P, the electron lines will diverge again and finally con verge to one point on the screen. This convergence is necessary in order to obtain a picture that is not hazy but sharp in all details. If the electrons do not converge to one point on the screen, it is possible to make the picture sharp again by altering the voltage of the first anode by means of a 'focus' control.

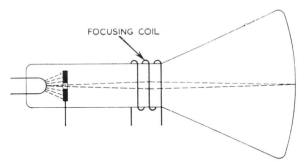

FOCUSING COIL

Figure 1.27 Electron beam focused to one point of the screen

In some types of tube (the magnetic c.r.t.), focusing is accomplished by means of a coil consisting of many turns of wire wound on a soft-iron ring provided with an annular air gap, as shown in *Figure 1.27*. An adjustable direct current in the winding sets up a strong magnetic field through the centre of the coil. Electrons moving exactly along the axis of the tube pass through the focusing field with no deflection, since they move parallel to the magnetic lines of force at all times. All other electrons diverge as they leave the cathode region because of their initial outward radial components of velocity. As these electrons intersect magnetic lines of force they are subjected to forces that have a focusing effect at the screen.

The current in the focusing coil can be varied by the 'focus' control. If this current is too high, the electron rays will take the form of *Figure*

Figure 1.28 Incorrect focusing: (a) current in focusing coils is too high, (b) current too low

1.28(a), and if the current is too low that of *Figure 1.28(b)*. In both cases the picture is dim and indistinct.

Focusing must be re-adjusted from time to time. By analogy with optical lenses for focusing rays of light, one speaks about electron lenses and electron optics.

Brilliance

If the grid is made sufficiently negative with respect to the cathode (e.g. −40 V), the electrons are repelled by the grid so that they cannot pass through the grid's hole. In that case electron ray and picture disappear. It is obvious that, by means of the grid potential, the brilliance of the picture can be adjusted from zero to a certain maximum. The control changing the grid potential is usually marked 'brilliance', 'brightness' or 'intensity'. Thus the grid has a double function: it converges the electrons to point P in *Figure 1.26(b)*, and also adjusts the electron current, i.e. the brilliance of the picture.

If there is a defect, the luminous spot should not be allowed to remain on one point of the screen as this may cause the screen to be burnt and permanently damaged.

Electrostatic deflection

From *Figure 1.25* it can be seen that the ray passes between two deflecting plates. If a voltage is applied between these plates (making, for example, the upper plate positive and the lower plate negative) the ray is deflected upwards, for the positive plate attracts the electrons (unlike charges), whereas the negative plate repels them (like charges). The very small mass of an electron means that it reacts at once to an electrical force.

The higher the voltage between the plates the greater will be the deflection, and the larger the deflection a of the luminous spot from the centre of the screen. It will be clear that, by reversing the voltage between the two plates, the ray is moved downwards, and also that an alternating voltage causes the ray to move up and down. If the frequency of this voltage is very low, it is possible to follow the motion of the luminous spot on the screen, but with a higher frequency, owing to the slowness of the eye and especially the afterglow of the screen, only a straight line is visible.

In the tube of *Figure 1.25* there is a second pair of deflecting plates, perpendicular to the first. If a potential is applied between them the ray is deflected in a plane perpendicular to that caused by the first plates, i.e. horizontally.

The apparatus of which the cathode-ray tube forms a part always contains a *sawtooth oscillator*. This is a generator of an alternating voltage that does not follow a sine law but produces the sawtooth waveform shown in *Figure 1.29*. A cycle lasts from A to B. This voltage is applied

Figure 1.29 Sawtooth voltage, which deflects the luminous spot of a c.r.t. to describe the timebase

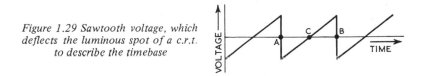

between the two deflecting plates that cause a horizontal deflection and that we will therefore call horizontal deflecting plates. (In fact the plates are in a vertical position.)

During the cycle from A to B, the spot will move on the screen from left to right (from right to left would be possible too). At the moments A and B the voltage suddenly changes from maximum positive to maximum negative, and consequently the luminous spot will jump from one end of its horizontal path to the other before travelling back again. At the moment C the voltage is zero, and the luminous spot is in the centre of the screen.

Figure 1.30 Timebase of a c.r.t.

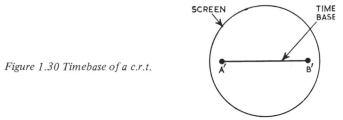

Suppose the peak voltage at the moments A, B, etc. is increased but the length of the cycle A-B remains the same. Then at the moments A, B, etc. the luminous spot will show a greater deflection; consequently it will move with increased velocity between these two moments. The line A'B' formed by the moving spot on the screen is called the *timebase* (see *Figure 1.30*).

To understand the working and possibilities of a cathode-ray tube in connection with applications described later, let us apply simultaneously a sawtooth voltage between the horizontal deflecting plates and a sine-wave alternating voltage between the vertical deflecting plates. One of the following three conditions may then occur:

1. The frequency of the sine-wave voltage is equal to that of the saw-tooth voltage (the sweep frequency), i.e. the duration of one cycle of both voltages is the same. Consequently, in the same time that (owing to the sawtooth voltage) the luminous spot moved from extreme left to extreme right, it will also move up and down once (owing to the sine-wave voltage). The result is that the spot traces out one cycle of the alternating voltage. The same occurs during the next cycle and the spot follows *the same path*. As this takes place again and again we see a fixed

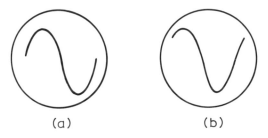

(a)   (b)

Figure 1.31 In (a) the sine-wave voltage is zero when the luminous spot is at the extreme left; in (b) this is not the case

image (*Figure 1.31(a)*). In this figure it has been assumed that, when the sine-wave voltage is zero, the luminous spot is at the beginning of the timebase sweep. This need not be the case: we could be confronted with the image of *Figure 1.31(b)*, for example.

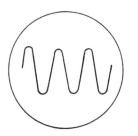

Figure 1.32 The frequency of the sine-wave voltage is an exact multiple (treble in this case) of the timebase frequency

2. The frequency of the sine-wave is an exact multiple (e.g. treble) of the frequency of the sawtooth voltage. This is represented by *Figure 1.32*.
3. The frequency of the sine-wave voltage is not an exact multiple of that of the sawtooth voltage. In this case, more or less than a whole number

of cycles of the sine-wave voltage will take place during the time that the spot follows the timebase once, e.g. the part a-b in *Figure 1.33*. The spot jumps back from point b to point c and then follows the path c-d, followed by e-f and so on. Each path will differ from the preceding one, and as there are usually a great many images per second the screen of the cathode-ray tube is sometimes dimly lighted all over. If the path is

Figure 1.33 If the frequency of the sine-wave voltage bears no exact relation to the timebase frequency, the sine-wave moves to the left or right with every sweep

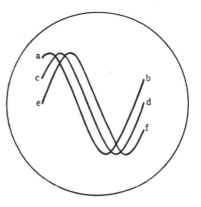

shifted only a little each time, the sine-wave curve will be seen moving to the left or to the right. A film of a moving ball is an analogy: each of the successive images represents the ball at a slightly further stage in its path, but the discontinuous movement of the ball is perceived as continuous on the screen.

The sweep frequency can be varied within very wide limits, and in this way any periodic voltage can be rendered visible as a fixed image. To do this it is necessary for the sweep frequency to be made equal to the frequency of the alternating voltage that we want to render visible, or for the latter to be a multiple of the former.

It is possible to render invisible the 'fly-back' of, for example, b-c in *Figure 1.33*: if a sufficiently high negative voltage is automatically applied to the grid during the fly-back, the electron ray is suppressed.

Measurement of time

The frequency of the sawtooth voltage can easily be made so high that the luminous spot describes the timebase in, for instance, 1/10 000 s. If the length of the path is 100 mm, the luminous spot moves 1 mm in 1/100 × 1/10 000 s = 1 microsecond (1 μs). With the help of a graduated scale on the screen, therefore, we can make visible on the screen a periodic voltage (or some other periodic phenomenon that has been converted into a proportional electric voltage) and see how it changes every microsecond.

Horizontal and vertical deflection by alternating voltages

Alternating voltages can be applied between both the horizontal and the vertical deflecting plates. The voltage between the horizontal plates causes the luminous spot on the screen to move up and down, and simultaneously the other voltage between the vertical plates causes the spot to move to and fro.

If the two alternating voltages not only have the same frequency but are also in phase, they will be zero at the same time. The spot then appears in the centre of the screen. After this point of time the deflections in the horizontal and vertical directions will be at any time proportional to the

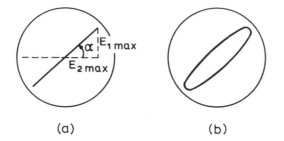

(a)   (b)

Figure 1.34 Two sine-wave voltages of the same frequency applied between the horizontal and the vertical deflecting plates cause (a) a straight line on the screen if they are in phase, (b) an ellipse if they are out of phase

instantaneous values of the two voltages. Hence the luminous spot follows a straight line on the screen; see *Figure 1.34(a)*. It is clear that:

$$\tan \alpha = \frac{E_{1\ max}}{E_{2\ max}} = \frac{max.\ deflection\ in\ vertical\ direction}{max.\ deflection\ in\ horizontal\ direction}$$

$$= \text{ the ratio of the two alternating voltages}$$

If the two alternating voltages are not in phase, the vertical movement is out of phase with the horizontal movement, and an ellipse will be seen, as in *Figure 1.34(b)*. This phenomenon can sometimes be observed in direction finders.

Tubes with magnetic deflection

The cathode-ray tubes provided with deflecting plates as described above are used in the Loran apparatus, in some radio direction finders and occasionally in echo depth sounders. We will now describe the cathode-ray tube used with radars.

A basic law of electricity is that a force is exerted on a conductor in which a current is flowing if the conductor is placed in a magnetic field. The force has a direction perpendicular to the direction of the current in the conductor and also perpendicular to the direction of the magnetic field.

The moving electrons in the evacuated cathode-ray tube also constitute an electric current. It is true that this current does not flow through a conductor, but if a magnetic field is set up in the tube a force will be exerted (see *Figure 1.35*) that will deflect the electron beam; the amount of deflection will increase as the field gets stronger. In *Figure 1.36*, where the magnetic field is set up by the current in the coils, the luminous spot is

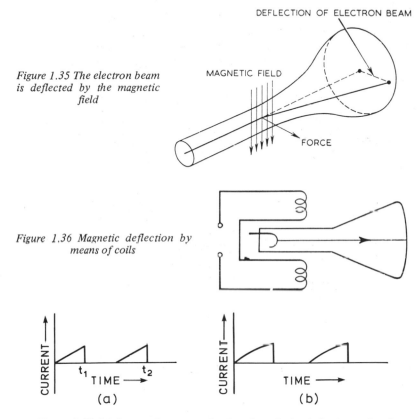

Figure 1.35 The electron beam is deflected by the magnetic field

Figure 1.36 Magnetic deflection by means of coils

Figure 1.37 (a) Sawtooth current flowing through the deflecting coils of Figure 1.36; (b) if the current does not increase linearly the image is distorted

deflected in a direction perpendicular to the plane of the diagram. This deflection is proportional to the current passing through the coil. If a current like that shown in *Figure 1.37(a)* flows through the coils, the luminous spot moves at a constant speed from the centre to the circumference of the screen, then falls quickly back to the centre to repeat the same movement a short time later. At the moments t_1, t_2 etc., when the current suddenly decreases from its maximum to zero, the deflection will also change from its maximum to zero. If the increase in current in the same time is smaller, the deflection of the luminous spot will be smaller, and the spot will therefore be moving slower.

The waveform of the current is not always perfectly straight, but may be as shown in *Figure 1.37(b)*. As the current increases rather faster initially than it does at the end, the speed of the luminous spot is somewhat greater at the start of the path and somewhat smaller at the end than it is in the middle. The consequence of a small non-linearity can be important, as the image appearing on the radar screen will be distorted (see also Chapter 9). It is, however, possible to adjust the current in such a manner that the non-linearity is reduced to a very small value.

Since the current in the coils causes deflection of the luminous spot, the coils are called *deflecting coils*.

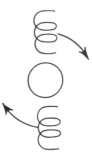

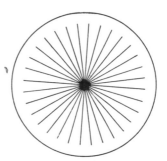

Figure 1.38 Deflecting coils rotating at constant speed round the neck of the tube

Figure 1.39 The image seen on the screen if the current of Figure 1.37(a) is fed to the rotating coils of Figure 1.38

It has been explained that the sawtooth current will cause a line to be described on the screen from the centre in a direction perpendicular to the axis of the coils. If the deflecting coils that carry the sawtooth current turn at a constant speed around the neck of the tube, as shown in *Figure 1.38*, the timebase turns around too (see *Figure 1.39*).

If the frequency of the sawtooth current is high enough and the afterglow of the screen is sufficiently prolonged, there will be so many sweeps during one revolution (perhaps 2000) that a lighted circular area will be seen. We will see in Chapter 9 how a radar image of the surroundings can be obtained.

It is clear that the sawtooth current must be fed through two slip-rings with brushes; the deflecting coils lie arched around the neck of the tube. The rotating magnetic field can also be obtained by mounting three fixed coils round the neck of the tube and passing through them currents having an appropriate waveform; the slip-rings are not required then.

A deflection can therefore be effected by means of a current (*electromagnetic* tubes) as well as by means of a voltage (*electrostatic* tubes). In electromagnetic cathode-ray tubes, the Aquadag coating, which may have a voltage of about 4500 V, is connected inside the tube with the screen.

THE HYPERBOLA

A hyperbola is the locus of all points in a flat plane that have a constant difference of distance from two fixed points. For instance, in *Figure 1.40*, the point C is 150 kilometres from A and 190 from B; the difference is

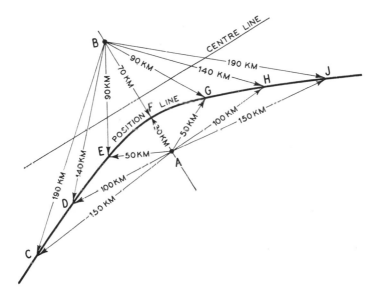

Figure 1.40 Hyperbola CJ for 40 km difference of distance from two fixed points A and B; AB is the baseline

therefore 40 km. The point D is 100 km from A and 140 from B; so for this point also the difference is 40 km. The same holds for the points E, F, G, H and J. If a curve is drawn through all the points that fulfil this condition, the result is the hyperbola for a 40-km difference of distance from A and B. Other hyperbolas could be drawn for other differences. It is obvious that a hyperbola can be described through any point of the plane.

In *Figure 1.40* AB is called the *baseline*, the extensions of it are the *baseline extensions*, and the perpendicular across the centre of the baseline is the *centreline*. The centreline and the baseline extensions are also hyperbolas (which have degenerated into straight lines) for differences of

distance of respectively zero and a length equal to the baseline. The fixed
points A and B are called the *foci*.

Figure 1.41 shows a number of hyperbolas for which the differences of
distance from the foci increase regularly from one to the next; in this case
by 20 km. As its distance from the baseline increases, each hyperbola
comes nearer to a straight line through the mid-point of the baseline. If the
hyperbola is extended infinitely, it eventually coincides with this straight

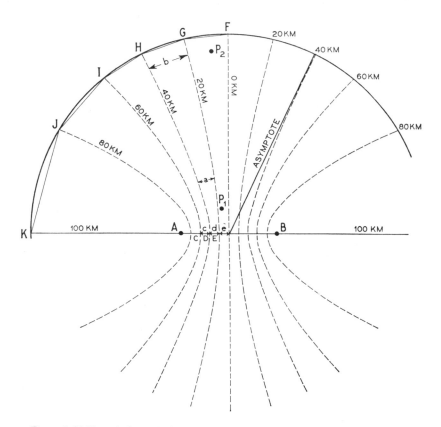

Figure 1.41 Hyperbolas with foci A and B; b > a, c = d = e etc., FG < GH < HI etc

line, which is called the *asymptote* of the hyperbola. In *Figure 1.41* the
asymptote of a hyperbola for a difference of 40 km has been drawn.
It should be noted further from *Figure 1.41* that:
1. The distance between two hyperbolas increases as the distance from
   the baseline becomes greater ($b > a$).
2. The intervals at which the successive hyperbolas cross the baseline are
   equal ($c = d = e$ and so on). This can be proved: C lies on the hyperbola

for a difference of distance of 60 km; therefore BC — AC = 60; (BD + c) — (AD — c) = 60; BD — AD + 2c = 60. As D lies on the hyperbola for a difference of 40 km, BD — AD = 40. Thus it follows that 2c = 20, and c = 10 km. It can be shown similarly that d and e = 10 km.

3. The distance FG < GH < HI, etc.

A hyperbola is a curve formed by the section of a circular cone and a flat plane when the plane cutting the cone forms a smaller angle with the axis of the cone than half the top angle.

Application to position-fixing

The basis of some position-fixing systems that will be described later is that it is possible to measure the difference of the distances to two transmitting stations A and B. If we find for this difference of distance a certain value, e.g. 20 km, we are on one of the two hyperbolas for this difference of distance (*Figure 1.41*). The foci of these hyperbolas are the transmitting stations A and B.

The differences of distance are measured on the surface of the Earth, taking into account the shape of the Earth. If the shape of the Earth is considered to be a sphere the lines of constant difference in distance are called 'spherical hyperbolas'. If the shape is considered to be an ellipsoid the lines are called 'spheroidal hyperbolas'. The lines are shown on charts.

In principle, the difference in distance is deduced from the difference in the time taken by radio waves emitted simultaneously by two transmitting stations A and B to reach the receiving ship.

Whether a ship moves from F to G in *Figure 1.41* or sails the distance e on the baseline, the increase in the distance-difference reading in both cases is 20 km. Suppose the accuracy to be 1 per cent of the reading; this 1 per cent represents a greater distance as the distance between the two hyperbolas increases (for instance $b > a$). So at P_2 the accuracy is less than at P_1. It is therefore desirable that the distance between two adjacent hyperbolas should increase as little as possible with distance from the baseline. This can be achieved by *making the baseline longer*. See *Figure 1.42*, where the ratio of the baselines in *(a)* and *(b)* is 1:3 but the widths of the lanes measured along the baseline are the same.

On sea charts, time differences for hyperbolic position-fixing systems are indicated in microseconds or other units near the hyperbolas. This enables the navigator to determine for a certain region the distance in nautical miles corresponding to a change of, say, 1 microsecond in the time difference. For this he only has to divide the number of nautical miles between the two adjacent hyperbolas near the position of his ship by the difference in the microsecond values marked near the hyperbolas.

It can be proved that the locus of all points with an equal accuracy is a circle through the two foci; see *Figure 1.43*. The number of nautical

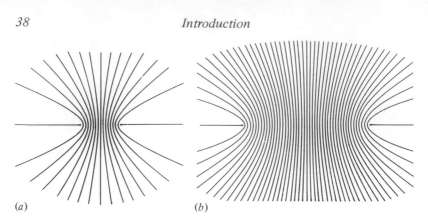

(a)   (b)

Figure 1.42 Because of the longer baseline in (b) than in (a), the distance between two adjacent hyperbolas increases less at a greater distance from the baseline

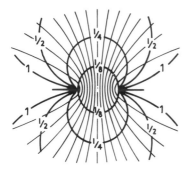

Figure 1.43 The locus of points with equal accuracy is a circle through the two foci; the geometrical accuracy (n. miles/μs) is indicated, for a given baseline, on each circle

miles corresponding to an error of 1 microsecond in the measured time-difference, for a certain baselength, is indicated for each circle.

REFLECTION BY IONOSPHERE

At a great height above the surface of the Earth, there are some ionized layers of air that play a big part in the propagation of radio waves. Some of the atoms of the component parts of the air are split into positive particles (ions) and negative particles (electrons) by the very high energy that the ultra-violet radiation from the Sun possesses at this height.

Just as metals are made conductive by the presence of liberated electrons, so are these ionized layers in the atmosphere. When the radio waves of a transmitter hit these layers, the liberated electrons are. set vibrating; this means that electric currents flow in the layers of air, resulting in reflection, or rather in a refraction or bending of the direction of the radiation (*Figure 1.44*). The vibrating electrons continually collide with the ions and this causes a loss of energy for the radiation.

There are two principal layers, the E-layer (or Kennelly-Heaviside layer) at a height of about 110 km, and the F-layer (or Appleton layer) at about 200–300 km. The part of the atmosphere in which these layers are found is called the *ionosphere*.

As the ionization is mainly due to the action of sunlight, it is evident that the layers are ionized more during the day than they are at night. At night there is deionization when the positive ions unite again with the negative electrons to form neutral atoms.

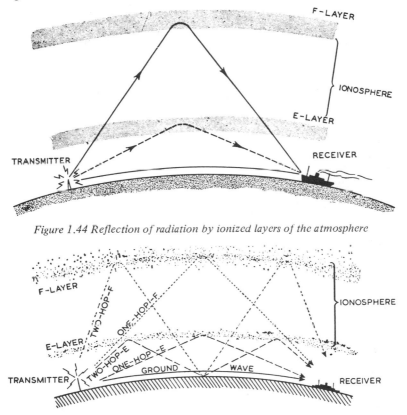

Figure 1.44 Reflection of radiation by ionized layers of the atmosphere

Figure 1.45 One very short transmitted pulse can be received as five pulses

The amount of bending of the radio waves by the layers, and the loss of energy involved, depend among other things on the frequency and the degree of ionization. The frequencies in use with the position-fixing systems to be dealt with later are mainly reflected by the E-layer. Frequencies above about 30 MHz (wavelengths shorter than about 10 m) are not reflected, but pass through all layers; this explains the limited range of these frequencies in communication between two points on Earth.

The radiation reflected downwards by the ionosphere is reflected again by the Earth, and this process is repeated until the energy has been expended (*Figure 1.45*). Therefore because of the ionosphere it is possible to receive transmissions from a very great distance.

The direct radiation, emitted in a horizontal direction, follows more or less the curvature of the Earth (the *ground wave*), but the loss when travelling along the Earth's surface is an important factor. The lower the frequency is, the farther the radio waves travel along the Earth's surface. At $f = 100$ kHz this distance is more than about 1200 miles, and at $f = 150$ MHz about 25 miles. The reflected waves (the *sky waves*) suffer loss of energy in the ionosphere but much less so, so that over large distances the sky waves dominate. For the frequencies in use with the position-fixing systems to be dealt with later, at an increasing distance from the transmitter one receives first exclusively the ground wave, then the ground and sky waves and thereafter only the sky waves.

During the day, when the layers are more ionized than during the night, there is more loss of energy. Sky waves are therefore received to a smaller extent during the day or not at all.

The range depends on the nature of the Earth's surface between the transmitter and the receiver. A rocky or sandy soil tends to reduce the range; sea water and a humid soil have a beneficial influence on the propagation and thus on the range.

Fading

If the ground wave and the sky wave are received simultaneously from the same transmitter, the two voltages in the receiving aerial are not in phase

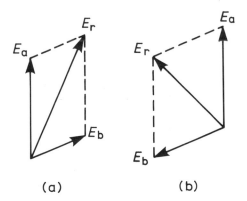

Figure 1.46 (a) Two alternating voltages, one received as a ground wave and the other as a sky wave, produce the resultant E_r; (b) the phase and magnitude of E_r change as a consequence of variations in ionization and altitude of ionospheric layers

(a)   (b)

because the sky wave has travelled a longer path than the ground wave. The same applies if two or more sky waves are received. See E_a and E_b in *Figure 1.46*, where E_r is the resulting voltage.

As the height and the ionization of reflecting layers are subject to variations, the paths covered by the waves will vary too. One consequence is that the *phase* of the resulting aerial voltage will vary. A second effect is that the *magnitude* of the resulting voltage will vary (see *Figure 1.46*); this phenomenon is called *fading*. The voltage can even be reduced to zero or to a level below the noise, making reception impossible. Both effects can have a very unfavourable influence on radio navigation systems, although if the receiver is equipped with automatic volume control (or automatic gain control) fading is hardly noticed.

NAVIGATION POSITION-FIXING BY HYPERBOLIC SYSTEMS

A new definition had to be found for the word 'navigation' to cover navigation in aircraft and spacecraft as well as on ships. The best definition seems to be that of Wing Commander E.W. Anderson: 'Navigation is the purposeful control of motion from place to place'.

By measuring the difference of the distances to two fixed points by one of the position-fixing systems to be described, the navigator is furnished, as we have seen, with a line of position of a hyperbolic character. A similar measurement to two other fixed points gives a second line of position, and the point of intersection of the two lines is the position of the navigator. The second line of position can, of course, also be found in other ways and may be an astronomical position line, a radio bearing, a depth line or a distance circle (determined, for instance, by means of radar).

By making use of *three* lines of position, the accuracy of the fix is increased. As each of the lines of position may have an error, these three lines will in general not intersect at one point but will form a triangle.

Fixed errors

Each measurement effected has an error, which consists of a fixed or systematic error and a variable or random error. *Fixed* errors are those that have a constant value and thus can be eliminated by correction; far from the coast the only way to measure fixed errors accurately is by means of a satellite-based position fixing system. All other errors are termed *variable*.

Sky waves have an inconstant and unfavourable influence on accuracy. During the day these waves are received less strongly or not at all. For the purpose of determining fixed errors we therefore take as the position found by the position-fixing system that which is determined *by day*. Hence, the fixed errors in radio position-fixing systems may be defined as the difference between the average of a large number of readings *by day* at a given position and the chart coordinates for the same position.

Fixed errors may be caused by a speed of propagation that differs from the speed assumed in computing the position lines on the chart. The effective speed of propagation varies widely with the electrical characteristics of the medium over which the signals pass. For example, the sum of experience so far points to a mean speed of 299 650 km/s over seawater transmission paths, while a corresponding figure for land paths of the lowest soil conductivity amounts to about 297 000 km/s.

For the computation of Omega charts a velocity of 300 574 km/s is taken. In as much as it is faster than the speed of light, it is worth noting that such a figure can only exist in a complex propagation system such as that of Omega signals in the waveguide formed by the Earth and the ionosphere.

Charts with hyperbolic lines can now be produced very quickly by automated techniques. For Decca charts the published corrections are compiled by computations of the theoretical chart readings from accurate geographical positions in the same geodetic survey network as that of the Decca transmitting stations; hence any small errors that might be contained in the chart itself (e.g. registration errors, position of coast lines, etc.) are not eliminated by the application of the published fixed-error corrections.

The sky wave follows a longer route than the ground wave. If, therefore, we receive the sky wave from both transmitters of a hyperbolic system, the result of the measurement will generally not be the same as that given by the ground wave. On the supposition that the height and the degree of ionization of the ionosphere do not change, the resulting error is constant for a given place. With the Loran system the sky wave is often received, and a correction is then applied based on the normal ionization and height of the ionosphere.

Variable errors

Variable errors are due to various accidental causes and the chances are equal that they will be positive or negative. We cannot, of course, apply any corrections for the variable errors, but we can take them into account. Suppose that a great many measurements have been made and that the error of each measurement has been determined by comparison with the exact value. We should be inclined then to use the largest of these errors as a measurement of the accuracy. This is misleading, however, because an extra-large error would give a wrong impression of the accuracy. Considering the great number of observations, we may take it for granted that in a few cases the error is smaller than the largest. We do not, therefore, use the maximum error, but mostly the 95 per cent error. This is *the error that is exceeded in only 5 per cent of all cases*. The 50 per cent error is also sometimes used. According to probability theory, the number of observations with a large error is much smaller than the number with a slight error, and consequently the 50 per cent error is three times smaller than the 95 per cent error.

If the value of the 95 per cent error of a line of position is known, it is possible to draw parallels on either side of this line at a distance equal to the error. It can then be assumed that the position of the observer is between these two parallels. For the other line of position, two parallels can be drawn in the same way. The size of the parallelogram so formed (shown shaded in *Figure 1.47*) is a measure of the accuracy.

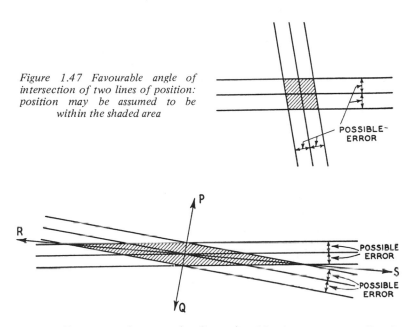

Figure 1.47 Favourable angle of intersection of two lines of position: position may be assumed to be within the shaded area

Figure 1.48 Accuracy decreases when lines of position intersect at a small angle

If the two lines of position intersect at a small angle (as in *Figure 1.48*) the parallelogram will become larger and so the accuracy will be less. It should be noted that the accuracy is still good in the direction PQ but is worse in the direction RS. The most favourable angle of intersection for the lines of position is therefore a right-angle.

Calculation shows that the figure inside which the position can be assumed to lie is really an ellipse rather than a parallelogram.

Accuracy of radio position-fixing

As proved above, the accuracy of a hyperbolic system is highest on the baseline. In contrast, however, the accuracy is very poor in the vicinity of the baseline extensions, and hyperbolic systems should not be used there, except under special circumstances.

The sites of the stations are chosen to give the greatest degree of accuracy in areas where it is most necessary; for example where there is a heavy concentration of ships. The ability to do this is an advantage of radio position-fixing compared to astronomical position-fixing, where the degree of accuracy is everywhere the same.

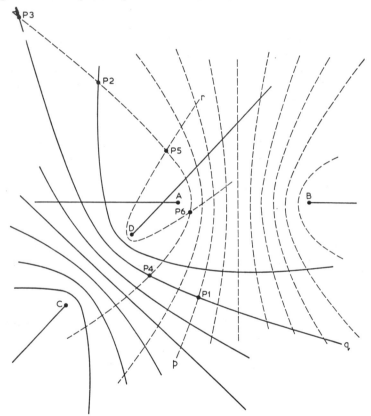

Figure 1.49 Position-fixing by hyperbolic systems. A and B, C and D are pairs of stations. The accuracy of P_1 is greater than that of P_2; the ambiguity between P_3 and P_4 can be solved, but not that between P_5 and P_6

In *Figure 1.49* two patterns of hyperbolas are shown. If we are on the line of position p with respect to the pair of stations AB, and also on the line of position q with respect to the pair of stations CD, our position is the point of intersection P_1. In this case, the angle of intersection is very good, and moreover we are not far from the two baselines, so a fair degree of accuracy may be expected here. This is not the case at P_2 where we are in the vicinity of the baseline extensions, and the angle of intersection is not so good.

Two hyperbolas may have *two* points of intersection (e.g. P_3 and P_4), which may give rise to ambiguity in locating the position, but in practice this will not usually cause difficulties, as the two possible positions are too far apart. If, however, under special circumstances it is necessary to make use of a line of position quite near to the baseline extension (e.g. the line r), the distance between the two possible positions (P_5 and P_6) may become so small that doubt arises. In such cases it will be possible to determine the correct position by noting the increase or decrease in the reading, taking into account the ship's course.

In the preceding pages we have considered the accuracy of radio position-fixing with respect to chart coordinates in longitude and latitude. It will be clear, however, that the chart and the fixed errors do not play any part if we avail ourselves *exclusively* of hyperbolic and other coordinates, where this is justified.

For instance, a fishing vessel may prevent damage to her nets by a wreck on the seabed if the hyperbolic coordinates of these wrecks are known. Therefore fishermen are particularly well-informed about the coordinates of wrecks on the bottom of, say, the North Sea. Variable errors, which change with time, are important in such cases but fixed errors are not. The ship must, of course, use for navigation the same system of coordinates as that in which the wrecks etc. are indicated.

As another example, if two naval ships want to meet at a *fixed time* at a position indicated by hyperbolic coordinates, neither fixed nor variable errors are of importance.

When the readings of the indicator of a position-fixing system are consistently the same at the same place, we speak of a good *repeatability*. If the readings vary, this will be due to the variable errors of the system or of the equipment. It is evident from experience that normally the *instru mental accuracy* is very high; an indication of this can be obtained by operating two or more receivers side by side, and comparing their readings.

Absolute and relative accuracy

By *absolute accuracy* we understand, for example, the accuracy with which a church spire is indicated in terms of latitude and longitude on a chart. The *relative accuracy* is the accuracy of this point with respect to other points in the same geodetic survey network.

Thus a complete archipelago remote from the continents, such as the Hawaiian Islands, might be shown with a fixed error on a world map, which would therefore be inaccurate in the *absolute* sense; if, however, each point on a chart of the Hawaiian Islands had the correct position with respect to other points on the same chart, this would represent a high *relative* accuracy.

The term 'reading accuracy' is often used, with reference to the accuracy of the readout itself. The reading accuracy of a voltmeter, for example, could be increased if we were to extend the pointer, enlarge the dial and sub-divide the scale. It is obvious, however, that this would result only in a suggested, not an *actual*, increase in the accuracy of the reading. The accuracy is, of course, determined for the most part by the possible error in the position of the pointer with respect to its scale, and this position has not been altered by enlarging the pointer or the scale. Any method of increasing the reading accuracy must therefore take account of the overall accuracy of the instrument if it is not to mislead the user.

Improving accuracy

In principle, there are two methods of increasing accuracy. The first is to erect on shore a receiving station for the radio position-fixing system concerned, by means of which we can determine the value of the overall error (fixed plus variable) at regular intervals. If this error is transmitted by radio to nearby ships, these can now correct the readings of their position-fixing equipment. This method can now be used, for instance, in exploration or other operations near the coast. In the future it will be used on a permanent basis in the Omega system for the navigation of ships up to about 200 n. miles offshore. The second method of improving the accuracy consists in taking a large number of observations at the same place and calculating the average value. If the observations are spread over a period the variable error will be reduced or eliminated. On a moving ship, the counterpart of this method is to plot position fixes at short intervals and estimate the mean track through the plots.

Chains

Normally, one of the stations of a hyperbolic system, the master, cooperates with two or three other stations, the slaves. Together the stations form a chain.

If there are two slaves, the three stations usually lie in an approximately straight line with the master station in the middle (a so-called *triad*). If the master operates with three slaves, the latter lie approximately at the points of an equilateral triangle with the master at its centre (a so-called *star*). A chain with four slaves is called a *square*.

A Loran transmitting aerial consists of a stayed mast, 190 or 400 m in height, with 24 or 6 aerial wires between the top and the ground (an *umbrella aerial*). A Decca transmitter aerial has a height of 50—100 m (see

Figure 1.50 Decca transmitting station (courtesy Decca Navigator Co. Ltd)

Figure 1.50). Because of the very low radiated frequency, some Omega transmitting aerials have very long aerial wires of more than one or two kilometres.

Figure 1.51 shows the calculated lines of equal accuracy A, B and C. The accuracy decreases from A to C, for two reasons:

1. The increase of the distance between two successive hyperbolas of the same master and slave.
2. The less favourable angle of intersection of hyperbolas.

   If all baselines of a chain were doubled in length, one would obtain a figure for accuracy similar to *Figure 1.51* but with double the scale. The area enclosed by, for instance, curve A then becomes $2^2 = 4$ times as large. In order to cover a large area with reasonable accuracy, the baseline should therefore be made as long as possible.

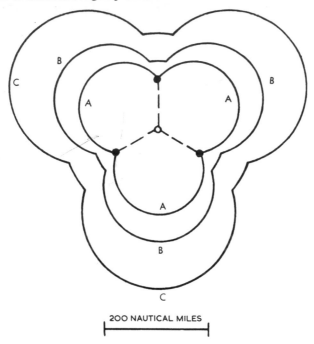

200 NAUTICAL MILES

Figure 1.51 Lines of equal geometric accuracy for a star chain

It will be appreciated that the accuracy of the fix is influenced by the synchronization of master and slave, as well as the accuracy of charts and corrections to be applied. It is possible, though in fact unlikely, for the radio position-fixing of, say, a point on a coastline to be more accurate than the position of the point as given on the chart.

Apart from the clarity of the visible horizon, the accuracy of an astronomical fix depends on the accuracy of the value of the apparent dip of the sea-horizon. Especially in the case of a calm, the correct value of the dip may differ considerably from the calculated one. One may set down, as a rule, that an astronomical fix is liable to a 95 per cent error of about 1.5 nautical miles.

Radio position-finding systems

There are more than twenty methods of determining one's position by means of radio. Several of them were originally developed for military applications and/or for civil aircraft. Some systems were abandoned after much research (and in some cases after enormous expenditure) because they could not meet the requirements or because better devices, techniques, or refinements of existing techniques were developed.

According to the position lines they provide, the systems can be divided into:

(*a*) *hyperbolic systems*, where the position lines are spherical or spheroidal hyperbolas; only these have been described so far;

(*b*) *radial systems*, where the position lines are radii (or, on a chart, great circles); examples are directional and non-directional (or omnidirectional) radio beacons, and the determination of direction by radar;

(*c*) *circular systems*, where the position lines are circles on the chart (range determination by radar, and the future system with atomic clocks as mentioned on page 217);

(*d*) systems giving other forms of position lines (e.g. satellite position-fixing).

2

Underwater navigational aids

THE ECHO SOUNDER

The echo sounder, an electrical device for measuring the depth of the water, has, since its appearance in about 1925, become such an important aid to navigation that it is now installed on almost every ship.

Principle

The principle of the measurement is as follows: in the water beneath the ship short pulses of sound vibrations are periodically produced, at a rate of 100 per minute, for example, and transmitted vertically to the bottom of

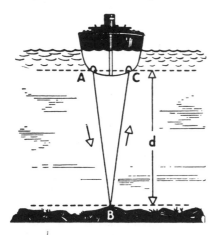

Figure 2.1 Principle of the echo soun-der. A short pulse of sound vibration is produced at A, reflected by the sea-bed at B, and received at C. The depth d is proportional to the mea-sured time interval between trans-mission and reception (courtesy Kelvin Hughes)

the sea. The sea bed reflects these pulses and, after a time that depends on the depth, the echo pulse is received back at the ship. In this time the pulses have traversed a path equal to twice the distance d between the keel of the ship and the bottom of the sea (*Figure 2.1*). If the water is not too shallow the 'Pythagoras error', due to the distance d being less than AB, may be ignored. The velocity of propagation of sound pulses in sea-water is practically constant and amounts to about 1500 m/s. Hence it follows that $2d = vt$ or $d = vt/2$, where v = velocity of sound in water and t = time interval.

For instance, if the time interval between transmission of the pulse and reception of the echo is one second, the pulses have travelled a distance of 1500 m in this time, and the depth is 750 m; if the time t = 0.5 s, the depth is 0.5 × 750 = 375 m. On a linear scale that indicates the time in

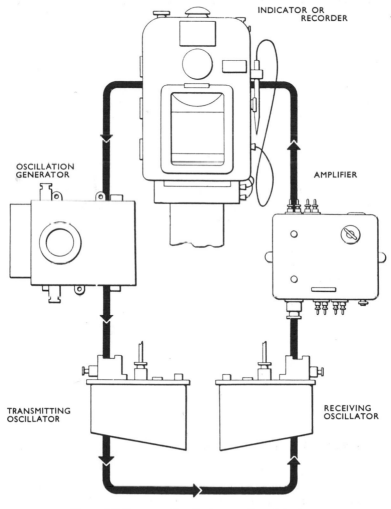

Figure 2.2 Components of echo-sounding apparatus

seconds, the corresponding depth in metres, fathoms or feet can be marked.

In shallow water, where the distance between the transmitter and receiver of the pulses cannot be neglected, the reading on a linear scale is greater than the true depth; in these circumstances it is therefore dangerous to neglect the Pythagoras error and to rely on the echo sounder.

The depth measured can be recorded automatically by apparatus that not only indicates the depth at any moment but also gives a survey of the depth along the route travelled. If, for instance, 100 soundings are taken per minute, and if the speed of the ship is ten nautical miles per hour, then the distance between two successive soundings will be about two fathoms, as can easily be calculated.

As a result of the short intervals, the navigator is warned at once of a sudden decrease in depth. The sea bottom near the coastline may have 'hills and valleys' like a range of dunes, and in course of time it may change considerably. In practice it is not always possible to indicate these changes immediately on the charts.

Apparatus

The apparatus consists of the following components (see *Figure 2.2*):
1. The oscillators. The transmitting oscillator starts vibrating when an electrical oscillation is supplied to it, while the receiving oscillator, which is set vibrating by the echo, converts the mechanical vibration back into an electrical oscillation. Oscillators are also called *transducers*. The vibrating surface of a transducer is in contact with the water, and is about 100 × 200 mm in area; see *Figure 2.3*.
2. The generator of electrical oscillations for the transmitting transducer.

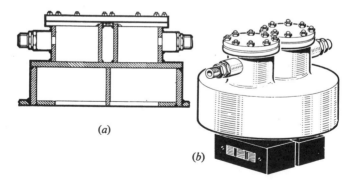

(a)

(b)

Figure 2.3 Transmitting and receiving transducers: (a) cross-section of the housing, (b) exterior view with the two transducers below (courtesy Kelvin Hughes)

3. The amplifier for strengthening the weak oscillations that the receiving transducer has converted from sound vibrations.
4. The indicator or recorder for measuring and indicating the depth. Items 3 and 4 are normally incorporated in one unit.

Reflection, refraction and absorption of sound vibrations in water

In *Figure 2.1* the sound waves to the sea-bed are indicated by a single line. In fact a beam is emitted (DOE in *Figure 2.4*) with a circular or elliptical cross-section.

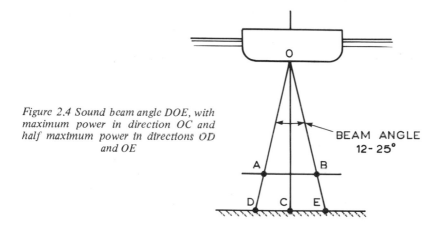

Figure 2.4 Sound beam angle DOE, with maximum power in direction OC and half maximum power in directions OD and OE

A circular beam has an angle between 12° and 25°. In *Figure 2.4* the intensity of the vibrations is not constant between A and B, nor does it suddenly become zero beyond A and B. In direction OC the intensity is greatest, and on each side it gradually decreases. The lines OD and OE are usually the directions in which the radiated power is half the maximum in direction OC. Just as in the case of a beam of radio waves with radar (see *Figure 9.33*) there are so-called 'side lobes' here that are not shown in *Figure 2.4*.

As the vibrations cross from one water layer to another of different composition or temperature, refraction and reflection occur. Suppose the vibrations cross from 'normal' sea water to a layer of plankton, a water layer of another salinity or another temperature (thermal layer), or a layer in which sand or other particles are suspended. In these layers the speed of sound differs from that in normal sea water; therefore at the separating surface there is refraction, with some reflection (usually weak unless the angle of incidence is a right angle). See *Figure 2.5*. The transition from one layer to another is usually not discontinuous, however, but gradual.

Instead of sudden refraction there is then a gradual bending of the direction of propagation.

The sea-bed is never completely flat, so the sound-waves are reflected in various directions (*diffuse reflection*) rather than in a single direction (*specular reflection*). In reality, a mixture of diffuse and specular reflection occurs. Diffuse reflection is required for an echo-sounder; for instance, if the reflection from a sloping sea-bed were exclusively specular, the ship would receive no echo. The shorter the wavelength in relation to the

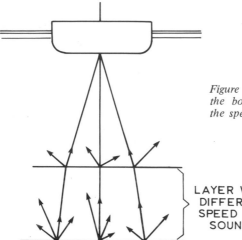

Figure 2.5 Reflection and refraction at the boundary of two layers in which the speed of sound is different. Diffuse reflection at sea-bed

LAYER WITH
DIFFERENT
SPEED OF
SOUND

average dimensions of the irregularities of the sea-bed, the less the specular reflection. So a shorter wavelength, i.e. a higher frequency, enhances the desired diffuse reflection. Another consideration is that during the time between pulse emission and echo reception the ship has moved somewhat; owing to the diffuse reflection, however, the transducer receives an echo nevertheless.

During propagation, of course, some absorption of the energy occurs, causing attenuation of the vibrations. In addition, as the beam cross-section increases, so the power per unit of cross-section decreases. Every doubling of depth increases the surface area of the beam four-fold, and thus makes the power per unit of surface four times smaller. Since this also applies to the echo, the power per unit of surface decreases (for this reason alone) by the fourth power of the depth.

Depending on the depth the power of the echo can therefore vary enormously. In order to restrict these variations, the amplification is automatically increased the later the echoes arrive (this is called 'swept gain'); the same method is used in radar receivers. Absorption can be decreased by decreasing the frequency. For very great depths of some kilometres, frequencies of about 10 kHz are therefore used. The echo-sounders for merchant vessels, however, have a maximum depth-range of about 1500 metres. Usually the minimum depth that can be measured is ½ to 1 metre and the frequency is about 40 kHz.

Speed of sound in water and frequency used

The speed of sound waves in sea water of normal salinity (3.4 per cent) and temperature (16°C) is constant and amounts to 1505 m/s. As a rule echo-sounders are adjusted to 1500 m/s.

It is clear that, if the speed of sound deviates from the velocity at which the apparatus has been adjusted, there will be a proportional error in the reading (called the 'speed error') so that the true depth = k times the indicated depth, where k is a proportionality factor that should be 1 at correct adjustment.

The speed of sound in water increases as water temperature, salinity and pressure increases; all these vary with depth. However, for normal applications on merchant vessels the deviations from the average value are small except for changing from salt water to fresh water. Assuming a correct adjustment for salt water, the true depth in fresh water is about three per cent less than the indicated depth.

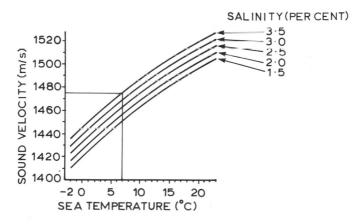

Figure 2.6 Relation between sea temperature and speed of sound in sea water, for various salinities

If necessary, corrections can be applied. *Figure 2.6* shows curves of sound speed versus temperature for several salinities. In general we may rely on the average speed of sound, since it is obvious that in shallow waters the deviations caused by abnormal temperature, salinity and pressure are small, and at great depths they are of less importance for the navigator.

As a result of increasing pressure (leaving aside the effect of temperature and salinity) the speed of sound increases by only about 1.8 m/s for
every 100 m of depth. Normally, however, temperature and salinity
variations have more influence, as depth increases, than the increase in
pressure. Sound speed can therefore increase or decrease as depth increases,
and in fact sometimes decreases in the top layers of the water and increases
again in the lower layers. *Figure 2.7* shows the speed of sound as a function
of depth, for a particular case.

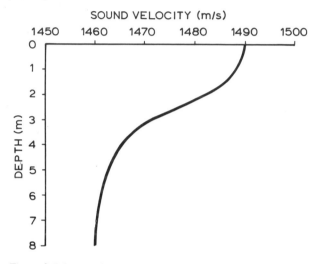

Figure 2.7 Speed of sound may increase or decrease with depth
as a result of changes in temperature, salinity and pressure

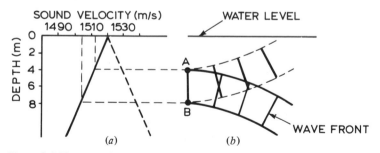

Figure 2.8 If sound velocity at A is greater than at B the beam will bend downwards; if it is less the beam will bend upwards

The *wavefront* of a beam is a plane perpendicular to the direction of
propagation of the sound. For some applications (for instance sonar) the
beam is directed horizontally, and consequently the wavefront is in a vertical plane. See *Figure 2.8*. If at point A the speed is greater than at B

(solid curve in (*a*)) the beam is deflected downwards (*b*) to greater depths. In the case of the dotted curve in (*a*), the speed increases with depth and so the beam bends upwards; when it arrives at the surface of the water, there will be practically complete reflection downwards again. However, with the echo sounder, bending of the beam does not play any part since the beam is directed vertically.

The number of pulses emitted every minute is between about 10 and 600, and the frequency of the sound is between about 10 kHz and 55 kHz (usually between 30 kHz and 50 kHz). The human ear can only detect sound vibrations with frequencies between about 30 Hz and 18 kHz. The frequencies used lie therefore as a rule beyond the audible range and are called *ultra-acoustic* or *ultrasonic* frequencies.

The main reason for using ultrasonic frequencies is that the screw and the ship's motion produce vibrations with frequencies in the audible range. When the receiving transducer has a resonance-frequency higher than this, these interfering vibrations will not be generally accepted or, if they are accepted to some extent, they cannot pass the resonance circuits in the receiver and so do not influence the reading.

A further advantage of using higher frequencies is that, for a given beam angle, the dimensions of the transducers can be smaller. On the other hand the loss of power at higher frequencies increases by the square of the frequency. The optimum frequency is, as mentioned, between 10 and 55 kHz.

TRANSDUCERS

Transducers can be divided into two kinds: *electrostrictive* and *magneto-strictive* transducers.

Electrostrictive transducers

Some crystal plates possess the property that, when a mechanical stress is applied to two opposite faces, electric charges of opposite sign are generated on them. If the compressing forces are changed to distending, the face that first was positive now becomes negative, and the other positive instead of negative (*Figure 2.9*). If a quartz plate is set vibrating mechanically, an alternating voltage is generated between the two faces. As is the case with so many physical phenomena, this property is reciprocal: when a voltage is applied between the two faces, the crystal contracts or expands a little, according to which face is made positive and which negative. The changes in the dimensions are so small that special methods are required to measure them.

This electrostrictive or piezostrictive property enables a crystal to be

used both for the transmission and the reception of ultrasonic vibrations in
water. It is clamped between two steel plates (*Figure 2.10*). When an alternating voltage is applied between the steel plates, the quartz and the plates
vibrate together. These vibrations would be very weak if no use were made

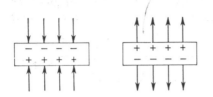

*Figure 2.9 When a quartz plate is
compressed, electric charges are generated on its faces; tension produces
charges of the opposite sign*

of the phenomenon of resonance. The frequency of the alternating voltage
is made equal to the natural frequency at which the system vibrates, and in
this way much stronger vibrations are obtained.

The lower of the two steel plates is in contact with the sea water and so
produces the vibrations in the water. The vibrations spread out mostly in a
direction perpendicular to the plate, which should therefore be fitted horizontally in the bottom of the ship. For crystals, quartz and, nowadays,

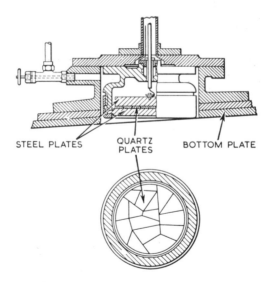

Figure 2.10 Electrostrictive transducer

more and more PZT are used, the latter being lead zirconate ($PbZrO_3$) and
lead titanate ($PbTiO_3$).

Often only one electrostrictive transducer, which serves as transmitter
as well as receiver, is installed on board.

Magnetostrictive transducers

When a bar of ferromagnetic material is subjected to a magnetic field (see
Figure 2.11) it undergoes a change in length, the value of which depends
among other things on the material of the bar and on the strength of the
magnetic field. This change may be an expansion or a contraction. As long

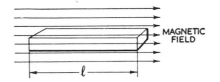

Figure 2.11 Bar of ferromagnetic material subjected to a magnetic field changes its length

as the bar is not pre-magnetized, it does not matter whether the lines of
force are in the direction shown in *Figure 2.11* or in the opposite direction. In both cases nickel, for instance, will contract.

If the length of the bar is l and the increase Δl (the decrease is consequently $-\Delta l$), the relative increase in length is $\Delta l/l$ and is very small.
Figure 2.12 gives the relative increase in length for some materials. This
property of magnetic materials is called magnetostriction, and it is displayed especially by nickel.

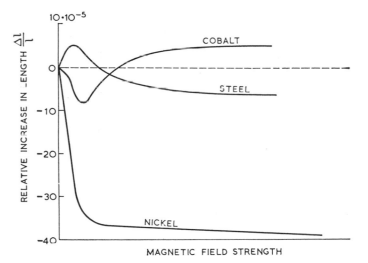

*Figure 2.12 Magnetostriction: relative increase in length of bars of
various materials placed in a magnetic field*

When a nickel bar is introduced into a coil (see *Figure 2.13*) in which an
alternating current flows, the length of the bar decreases and increases
periodically under the influence of the alternating magnetic field. Whenever the current and the field set up by it are zero, i.e. twice in every cycle,

the bar is its normal length, but when there is a current in one direction or the other it contracts. So the bar vibrates at a frequency which is twice that of the electric current. By giving the alternating current the frequency that makes the bar vibrate at its natural frequency, the vibrations become much stronger.

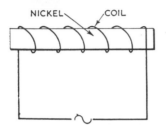

Figure 2.13 Magnetostrictive transducer: a nickel bar vibrates mechanically when an alternating current is supplied to the coil

The vibrations obtained in this way are transferred to the water in which the bar is placed. The bar should be correctly formed and installed in order to direct the vibrations as much as possible to the bottom of the sea.

The reverse phenomenon too may occur: changes in length of the bar produced mechanically will make the nickel magnetic, and on the bar being set vibrating, an alternating magnetic field will arise in the coil. As is known from the theory of electricity, this causes an alternating voltage to be set up in the windings of the coil.

Like iron, nickel has the property of residual magnetism, though to far less a degree. The nickel core of the receiving transducer is premagnetized; it has retained a small amount of the magnetism.

The echo that comes from the bottom of the sea causes the nickel of the receiving transducer to vibrate mechanically at its resonant frequency. Its residual magnetism will now alternately be strengthened and weakened, so that a voltage is induced in the windings of the coil of the same frequency as that of the vibrations in the water. The residual magnetism in the nickel of the receiving oscillator is used to obtain greater magnetic changes for a given mechanical vibration, so that the alternating voltage in the coil is also increased. As a rule, transmitting oscillators are not premagnetized.

Very powerful oscillators, used in special applications, can convert more than 1000 watts electrical power into sound. The same oscillator can convert sound into electrical power if only 0.000 000 05 watt sound is received.

Types of magnetostrictive transducer

There are two principal types of magnetostrictive transducer:
1. *Laminated nickel packs*. Thin nickel laminae insulated from each other

and having the form shown in *Figure 2.14* are piled up and joined together; when an alternating current is passed through the windings of the coil the nickel starts vibrating. Consequently each particle of nickel moves up and down, and as the upper side of the nickel is rigidly

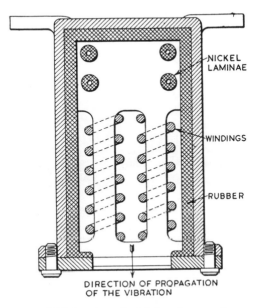

DIRECTION OF PROPAGATION
OF THE VIBRATION

*Figure 2.14 Laminated nickel-pack type of mag-
netostrictive transducer (courtesy Kelvin-Hughes)*

fixed, the lower side transmits the vibrations to the water with which it is in contact. The nickel pack is built up from thin insulated laminae to avoid eddy currents, which would cause a great waste of energy; the armatures of motors and dynamos are laminated for the same reason.

2. *Laminated nickel ring. Figure 2.15* shows another type, containing a ring of thin annular nickel laminae. The alternating current is passed through well-spaced and insulated windings of heavy-gauge wire. As the two diameters of the ring increase and decrease periodically, the outer surface will produce vibrations in the water. Only a reflector placed in the water can direct the acoustic energy in a definite direction (see *Figure 2.16*).

In 2, the vibrations are reflected not so much by the inner surface of the hollow double-walled reflector as by the contact surface of metal and air. This is not so strange if one considers a mirror, where it is not the glass surface that reflects the light, but the mercury behind the glass.

The following may serve to make this clear. The resistances that vibrations encounter during their propagation through various media are called acoustic resistances. The greater the difference between these, the stronger the reflection produced at the boundary between one medium and another. These resistances have been determined for various substances, so that we know the percentage of the energy that is reflected and the percentage that passes into the second medium. The latter is, for instance, about 10 per cent for sound vibrations that pass from water to iron, and only about 0.1 per cent for those that pass from water to air. This explains many

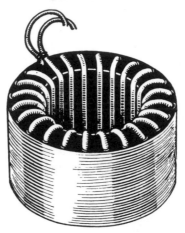

Figure 2.15 Laminated nickel-ring type of magnetostrictive transducer (courtesy Kelvin Hughes)

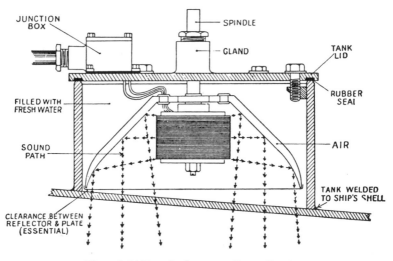

Figure 2.16 Use of reflector to direct vibrations

phenomena, such as the strong echo produced by small air bubbles in water, the nearly complete reflection against the water surface experienced by echoes coming from the sea bottom, and also the reflection by the layer of air in the hollow reflector mentioned above.

Though the vibrations produced by magnetostrictive transducers are amplified considerably by resonance, the changes in the dimensions of the nickel still remain so small as to be detectable only by very sensitive instruments. With a correctly constructed transducer these changes can, however, bring about very strong vibrations in an almost incompressible medium like water. Very strong vibrations will produce vacuum bubbles in the water, called *cavitation*. Since no mechanical vibrations can be transferred to a vacuum, this impedes the propagation of sound.

Mounting the transducer(s) in a tank has the great advantage that the ship need not be dry-docked when the transducer(s) must be repaired. The drawback is the already mentioned big loss that occurs when the vibrations pass through the hull plates. As the echo vibrations also suffer a loss, only a very small part of the transmitted energy will return.

When the hull plate through which the vibrations need to pass is too thick it is replaced by a thinner plate of rust-proof steel. In order to prevent the water in the tank from freezing when the vessel is dry-docked, an anti-freeze solution (e.g. glycol) is added to the water. Sometimes the tanks are filled with some other liquid than fresh water.

By mounting magnetostrictive or electrostrictive transducers in a *sea chest* the vibrations can be transferred directly to the water below the keel without passing through the hull plates. The use of a sea chest again makes it possible to remove and repair the transducer without dry-docking (see page 84).

Generation of electrical oscillations

As a rule one of the following methods is applied nowadays for the generation of electrical oscillations:

1. A capacitor is charged to a high voltage and then connected to the coil of the magnetostrictive transducer. The capacitor discharges through the coil, giving rise to a damped oscillation (*Figure 2.17*).
2. Normally, oscillations are generated by a tube or transistor oscillator and amplified in one or more stages. The advantage of this method is that the frequency, amplitude and duration of the oscillations can be adjusted more easily.

INDICATORS

Echo sounders can be divided into two kinds: *recorders* (echographs) and *echometers*. Echometers only indicate instantaneous depth; echographs also record depths. If the vessel moves at a constant speed, the record

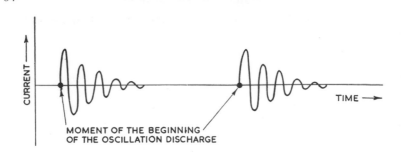

Figure 2.17 Damped oscillation

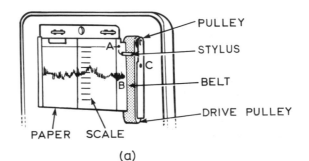

(a)

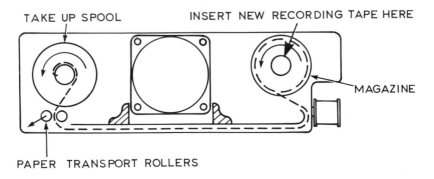

(b)

Figure 2.18 Simrad recorder. (a) Paper moves to left, stylus moves downwards; if pulse is transmitted at C instead of at A, the scale range is changed, say, from 0–50 metres to 30–80 metres. (b) Plan view: dotted line is correct paper travel

shows an automatically recorded profile of the sea-bed along the route of the ship.

Recorders

Recorders use a broad strip of paper that is drawn over a flat metal surface (the 'desk'), being wound slowly from one spool to another (*Figure 2.18*). A spool is sufficient for 30 to 100 hours; the approaching end of the spool is marked, for example, by a broad red line on the paper.

A belt on which the stylus is fastened runs over two pulleys, driven at a constant speed; thus the stylus trails down the front of the paper, returns to the top behind it, trails down the front again, etc. The paper is coated with a very thin metallic layer (e.g. aluminium powder), covered with a non-conductive layer. When the receiver supplies an electric voltage to the stylus, this burns away the upper layer of the paper to reveal the metallic layer, which is of a different colour. There also exists paper that reacts to heat. Another kind of paper (now very rarely used) contains iodine; an electric current, produced by the echo, causes a chemical reaction that gives rise to a brown discoloration.

When the stylus is at point A in *Figure 2.18(a)* it receives an electric voltage, which produces a dot on the paper. At the same time the emitted pulse starts travelling through the water to the sea-bed; its echo returns to the receiving transducer, where it is amplified and supplied to the stylus. As a result of this a second dot appears. The greater the depth the greater the distance between the two dots.

As the paper moves slowly to the left, the successive dots form two lines. The upper line is a horizontal straight line, the lower one a profile of the sea-bed. The depth can be read from a glass scale in front of the paper or from a scale printed on the paper.

Suppose that point B in *Figure 2.18(a)* corresponds to a depth of 40 metres and that the scale range is 0–50 metres. The pulse can also be emitted when the stylus is at C instead of A. (At C the stylus is on its way upwards behind the paper.) When an echo is recorded at B, the stylus has now moved from C to B instead of from A to B, representing a correspondingly greater depth; therefore a certain number of metres should be added to all scale readings, depending on the position of C. For example, if 30 metres were added to the readings, the scale range would then be 30–80 metres. In this way the apparatus can be adjusted to different 'phases'.

If we do not know the depth we should always start at the first phase, beginning at zero, and proceed to further phases if necessary.

It is clear that the velocity of the stylus should be constant. The electric motor driving the pulley should therefore run at a very steady number of revolutions per minute. For a correct indication this number should be (and should remain) adjusted for the standard speed of sound. (The latter is subject to alterations, but for normal applications these are neglected.)

For types with a fathom scale the pulley speed can be made six times greater; the same numbers on the scale will then indicate feet instead of fathoms (1 fathom = 6 feet).

Echometers

With a certain type of echometer, depth is indicated by a neon lamp moving at a constant speed behind a circular scale; see *Figure 2.19*. (Unlike

Figure 2.19 Echometer with neon-lamp indicator (courtesy Kelvin Hughes)

an incandescent lamp, a neon lamp lights/extinguishes immediately upon being switched on/off.) As the lamp passes the zero of the circular scale the pulse is transmitted, and when the echo arrives the lamp lights up for a very short while. The repeated illumination of the same point of the scale enables the navigator to read the depth, even from a distance.

On other makes of echometer the depth is indicated by a pointer or shown digitally. Some of these echometers have the feature of automatically warning the navigator, by an audible and/or visual signal, when the measured depth is greater or smaller than a pre-selected depth. A third type of echometer, for fishing craft, uses a cathode ray tube as indicator not only of the depth but also of the presence of fish in the beam.

Interval marker

In order to obtain a time scale on the paper, the 'interval marker' can supply an electric pulse to the stylus at intervals of, say, 3 minutes. This

Figure 2.20 Echogram of the New Waterway showing increased depth off Maassluis

results in dots on that side of the paper farthest from the zero line or on the other side of the zero line (*Figure 2.20*).

Draught setting

If the pulse is transmitted at the moment when the pointer or stylus passes the zero of the scale, the distance from the transducer to the sea bottom is read off. This is not exactly the same as the depth below the keel, the transducers not always being located at the lowest point of the vessel.

On some types it is possible to adjust the zero line with the 'Zero control' (also called the 'Draught setting'). To read off the real depth of the sea bottom, the distance from the water level to the transducers, measured vertically, should be added to the reading. In merchantmen, this distance may alter considerably. By means of the draught setting we can make the zero-line or transmission-line coincide with the figure on the scale that indicates the distance from the water level to the transducer.

There is always a risk, of course, that this adjustment might give a false feeling of security because the reading could be taken as the depth below the keel. But whatever the adjustment, allowance should be made for two facts:

1. With separate transducers, the transmitted pulse needs some time to travel directly via the hull or the water from the transmitting to the receiving transducer. This is especially the case when the distance between the two transducers is great. The transmission will not occur when the stylus passes the zero of the scale, but somewhat later.
2. The transducers may be at a higher level than the lowest point of the keel.

Checking the echo sounder

The true depth is equal to the sum of the depth indicated by the echo sounder and the vertical distance between sea level and the transducer(s). Not only the latter but also the indicated depth can change in the course of time. It is therefore advisable to check the echo sounder a few times a year, when an opportunity occurs, with the aid of a sounding by lead. This can be done, for instance, while waiting for a pilot to be taken on board. The sea bottom should be flat or slope gently. Two or more soundings by lead are then taken as close as possible to the transducer, to measure the true depth.

If the depth indicated by the echo sounder is not correct, the motor driving the stylus over the paper runs too fast or too slow. If the reading is too high the number of revolutions must be made smaller and vice versa. After correcting the speed of the motor (if this is possible) a new series of checks should be made with the lead.

In many types of echo sounder, the number of revolutions per minute of the stylus arm can be determined with the help of a tachometer. It is very important that it should be done after the motor has run for some time and the temperature has become constant, so that the speed no longer changes. Moreover, the voltage supplied by the mains should have the normal value.

For hydrographic work another, more accurate, method is often used for the calibration of the echo sounder, called the *bar check*. By means of vertical ropes, a tube filled with air and having a rectangular section is lowered in a horizontal position beneath the ship to a known depth. The reflection by the air in the tube causes its depth to be indicated by the echo sounder and by comparison any error is revealed. The advantage of this method is that measurements can be effected at any desired depth, as long as there is sufficient water under the keel. It can be used only on small craft, however.

Cross-noise

The number of controls for a very simple echo sounder is often only two or three, namely an on/off switch (with sometimes an intermediate position, 'standby'), a continuous or step-by-step adjustment of the amplification ('gain' or 'sensitivity'), and a switch or control for adjusting the illumination of the scale or the record ('dimmer'). The amplification control must be set to the position that produces the clearest echo-line.

If the sensivity is very great, it will be noticed that, just below the scale zero, a rather broad line or separate narrower lines accompanied by several irregular dashes and dots often appear (see *Figure 2.21*). This is called

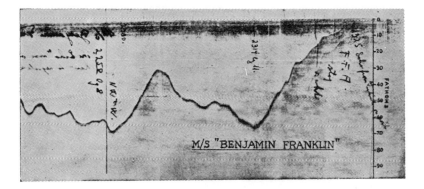

Figure 2.21 Echogram showing cross-noise

cross-noise, and is caused by a part of the vibration energy sent out, which by various ways (through the bottom plates, through the air and through the water) goes directly to the receiving oscillator. As the lengths of these paths are not equal, and especially as the velocities of propagation in the various media differ widely, each pulse emitted will be received and recorded as a series of pulses, as a result of which several lines may appear near the zero line or merge into one broad line.

Though excessive cross-noise is not desirable, some leakage is advantageous, because it can serve as an infallible indication of the instant of transmission. If excessive cross-noise occurs, however, the echo line can be totally or partially masked in shallow depths (see *Figure 2.21*). To avoid this, the amplification should be very small immediately after the moment of transmission, and increase gradually so as to record with sufficient clarity the weak echoes that might arrive from great depths. This amplification adjustment is carried out automatically; it is called 'swept gain', and should not be confused with the manual amplification control ('sensitivity' or 'gain').

Controls

The following switches and controls may also be found (see *Figures 2.22* and *2.23*):

(a) a range switch or 'phasing' control;

(b) if there is no metre scale, a switch to change from a foot to a fathom scale and vice versa;

(c) a 'fix marker' or 'event marker' push-button; when this is pressed, the stylus is supplied with an electric voltage and draws a line over the full width of the paper; this is done to mark a certain time, e.g. if bearings have been taken from a landmark;

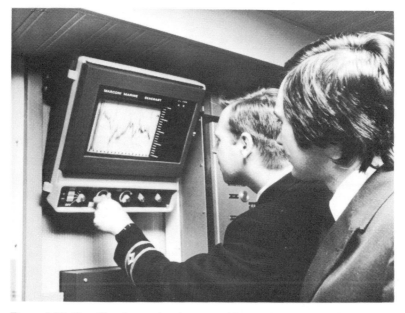

Figure 2.22 'Sea Chart' recorder (courtesy Marconi International Marine Co.)

(d) a control for changing the speed of the paper-transport;

(e) the zero-adjustment or draught setting control, mentioned before;

(f) a control to stretch the paper, since without a good electric contact between paper and desk no current can pass;

(g) a switch to change from a fore to an aft transducer or from a port to a starboard transducer (see page 75);

(h) a control for adjusting the minimum depth at which the alarm is to be given;

(i) a switch to change the pulse length and, at the same time, the number of pulses per minute;

(j) a 'normal/contour line' switch: at the setting 'contour' the recorder shows the bottom profile as a thin line, to facilitate the depth reading; at the 'normal' setting more details of the sea-bed are shown;

(k) a control to change the speed of the motor that drives the stylus over the paper;

In shallow waters switch (i) should be set to a short pulse (e.g. 0.3 milliseconds) because the pulse should terminate before its echo arrives, pulse

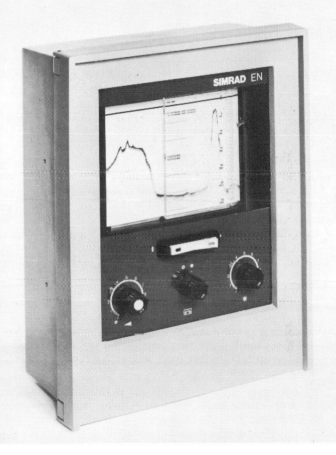

Figure 2.23 Type EN recorder (courtesy Simrad)

and echo taking an extremely short time to travel their path. When the apparatus is adjusted for greater depths, the number of pulses per minute may automatically be lowered, to avoid the echo of a pulse being received after the next pulse has already left the ship; this would lead to a large error in the reading. Simultaneously with a decrease of the number of pulses per minute the pulse length is automatically made longer (e.g. 2 milliseconds); thus the average power remains about the same.

By opening the front of the echograph it is possible to write notes on the paper. Depending on the type of paper, this should be done either with a ballpoint or pencil, or with an 'electrical pencil'; the latter is an insulated metal rod that is energized via a flex to produce the same effect as the stylus.

Second and third echoes

With a rocky sea bottom and maximum sensitivity, especially in shallow waters, there are often several echo lines visible (see *Figure 2.20*). This is due to the fact that the pulses are reflected a number of times up and down between the sea bottom and the keel or the surface of the water. In such cases only the first echo, of course, is taken into account. However, if the echo sounder has been adjusted to too great a range (e.g. 25–50 m), it may happen that not the first echo (say, 15 m) but only the second on 30 m and the third on 45 m are indicated, with the consequent risk that the second echo is taken for the first. The depth read would then be much greater than the true one. Always, therefore, begin by adjusting to the smallest range, and if no echo becomes visible, pass successively to the larger ones.

With many echographs each range starts at zero metres, thus eliminating the risk of confusing the second or third echo-line with the first one. Less detail of an echo-line at great depth appears then, however.

False soundings at great depths

At great depths it may occur that, in the time during which the pulse is on its way to the sea bottom and back, the stylus has made more than one complete revolution. The echo line then appears at, for instance, 75 metres, whereas the actual depth amounts to 75 metres plus the distance corresponding to one or even more revolutions. If this latter distance is 750 metres, the actual depth may be 75 or 825 or even 1575 metres. The British Hydrographic Office has received reports of uncharted shoals observed, which in most cases proved after an extensive and very expensive investigation to be erroneous.

Practical considerations

If the transducer is not contained in a tank, care should be taken when painting the bottom plates that the transducer is not painted or sand-blasted; this would impede the transfer of the vibrations to the water.

Moreover, painting is not necessary since the bottom plate of an electrostrictive transducer is made of stainless steel. In the case of a magnetostrictive transducer contained in a tank, part of the bottom plate is replaced by a thinner plate of stainless steel. Again, this plate should not be painted.

Before the ship leaves a dry-dock the stainless steel should be cleaned carefully (e.g. with thinner) in order to remove all oil, grease etc. The instructions of the manufacturer must be followed. It is the task of the mate to check, prior to de-docking, that the transducers are not painted.

Interpretation of echograms

On every echogram the scale for the depth is considerably larger than the scale for the time, perpendicular to it. A consequence of this is that a gently sloping 'valley' of the sea-bed is reproduced on the echogram as a steeply sloping cleft. See *Figure 2.20*, which shows the increased depth of the New Waterway off Maassluis.

The time scale on the echogram depends on the speed of the paper transport in the recorder. With many types this speed can be altered; the distances between the interval markers will then change. Paper can, of course, be saved by a lower paper speed.

Air bubbles in the water reflect sound vibrations particularly well. Hence, when the ship is going astern and air bubbles are conveyed under the keel by the action of the propeller, the echoes of these bubbles appear near the zero line.

There is the risk of a 'Pythagoras error' in shallow waters, as mentioned on page 52.

Compared with a bottom consisting of rock or gravel, a muddy bottom produces a weak, and therefore less clear, echo. A rocky bottom also gives rise to a second and third echo-line. If the reflection of rock and gravel is taken as 1, for coarse sand it is 0.03, for fine sand 0.01 and for mud only 0.001. Holes in a rocky bottom are often filled up with mud, and there may then be separate echo-lines from the mud-surface and the rock below it.

If the sea-bed is sloping (*Figure 2.24*) echoes arrive first from A, then from intermediate points and last from B. As a consequence of the shorter path, and hence smaller absorption, the echo from A should be strongest. However, the power radiated in the direction of A is less than in the direction of C. The strongest echo does not appear therefore from A but from somewhere between A and C. Thus the echo-line becomes broad and vague and the true depth is not indicated by the recorder.

It may occur that a narrow cleft or gully in the sea-bed (*Figure 2.25*) is not clearly recorded as such, because the echoes from A cannot be separated from those from B and C, which arrive earlier.

Vertical movements of the vessel, due to a considerable swell, can give a 'waved' course to the echo-line. Rolling and (to a smaller degree) pitching cause the direction of the beam to alter with the rhythm of the rolling. At a large angle of inclination the path covered becomes longer and the angle of incidence more unfavourable. Because of the longer path, variations may arise in the reading. In this case the echo-line again takes on a waved form. Vessels for hydrographic purposes, which need to measure depth very accurately, cannot operate under such circumstances.

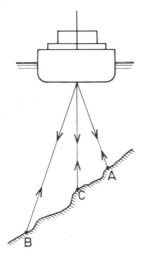

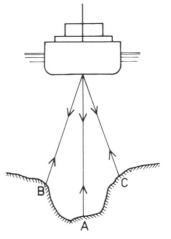

Figure 2.24 Sloping sea-bed: echo from A arrives first

Figure 2.25 Narrow gulley will not be clearly recorded

Neon-tube echometers do not show variations in light-intensity if the echoes become stronger or weaker.

For navigation purposes echographs are more suitable than echometers, because an echograph not only shows the present depth but also previous depths and details of them. When arriving at or leaving a port, as well as on the high seas, an accurate knowledge of the depth and its changes can provide an indication of the ship's position if detailed depth lines, and especially sudden changes in depth, are shown on the chart. Modern charts pay more attention to this.

Transducer configuration

The place where the transducers are mounted is very important and must be carefully chosen, especially on ships capable of high speed. It must be free from turbulence and undesired sound vibrations, and sufficiently far from the exhausts of pumps, etc.

The number of transducers may be limited to one, used for both transmission and reception. However, there are often two, both connected to the same indicator; one serves for transmission and the other for reception of the echo. The advantage of such *dual transducers* is that a smaller depth can be measured.

A transducer can be connected to two recorders or to an echometer and a recorder. In the latter case the echometer, with a small range, is installed on the bridge for instantaneous reading of the depth, while the recorder is installed in the chartroom.

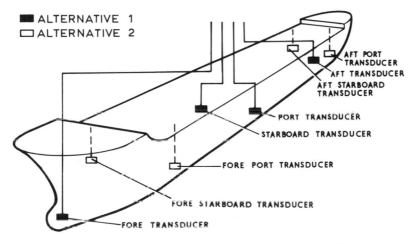

Figure 2.26 Two possible configurations of echo-sounders on very large ships

Large ships and passenger ships may be equipped with two transducers, one fore and one aft, each operating as both transmitter and receiver. The transducer to be used can be selected with a switch. The navigator can then be better informed in shallow waters about changes in depth. Also, it is possible that one of the transducers may function better under bad weather conditions than the other. If there are two echometers, or one echometer and a recorder, each of them may be connected to each of the two transducers.

Very large ships (VLCCs and LNGCs) may have either fore and aft dual transducers, or midships dual transducers and fore and aft single transducers. *Figure 2.26* shows these two alternatives.

THE DOPPLER LOG

The Doppler log is based on measurement of the Doppler effect. It has been proved (page 24) that an observer, moving with a source of sound towards a reflecting plane, receives a frequency:

$$f_v = f\frac{c+v}{c-v} \qquad (2.1)$$

where f_v is the received frequency, f the transmitted frequency, c the speed of sound and v the speed of the source of sound.

By measuring f_v and knowing f and c, the speed of a ship with regard to the sea-bed can be determined; this method can be applied not only to the alongships but also to the thwartships speed. By supplying these speeds to an integrator the distances covered in both directions can be calculated. Starting from a known position, the estimated position can be determined in this way at any moment.

Principle

A transmitting transducer below the ship continuously emits a beam of sound vibrations in the water at an angle α (usually 60° to the keel) in the forward direction; see *Figure 2.27*. A second transducer aboard receives the echo caused by diffuse reflection from the sea-bed.

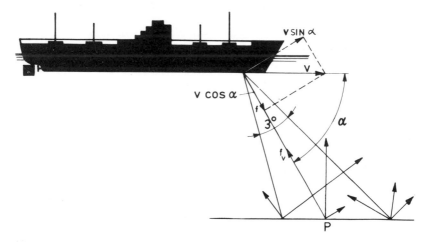

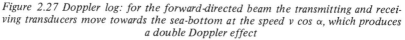

Figure 2.27 Doppler log: for the forward-directed beam the transmitting and receiving transducers move towards the sea-bottom at the speed v cos α, which produces a double Doppler effect

A Doppler log uses a higher frequency than an echo sounder. In the first place, the resulting shorter wavelength leads to the more diffuse reflection desired; the echo from a specular reflection would not be received, in view of the oblique incidence of the beam. Secondly, the shorter wavelength makes possible a smaller beam-angle and so avoids the dimensions of the radiating face of the transducer becoming too large. Thirdly, the emitted power of the sound vibrations spreads less and thus the echo is stronger.

In *Figure 2.27* we can replace the vector for the speed v of the vessel by two mutually perpendicular vectors, the direction of one of the vectors, v cos α, coinciding with the middle of the beam. Since one component of the ship's velocity (the vector v cos α) is in the direction of the beam, the received frequency f_v will differ, due to the Doppler effect, from the transmitted frequency. Equation 2.1 now becomes:

$$f_v = f \frac{c + v \cos \alpha}{c - v \cos \alpha} = f \frac{1 + (v/c) \cos \alpha}{1 - (v/c) \cos \alpha} \tag{2.2}$$

The other vector of the speed of the vessel, v sin α, is directed perpendicular to the beam and hence will not cause a Doppler effect.

It can be proved mathematically that for small values of x with regard to 1:

$$\frac{1}{1 - x} = 1 + x + x^2 + x^3 + \ldots$$

This is an infinite progression. Because the successive terms decrease very quickly, those smaller than x^3 may be neglected. In equation 2.2, (v/c) cos α is very small with regard to 1, so:

$$f_v = f \frac{1 + (v/c) \cos \alpha}{1 - (v/c) \cos \alpha}$$

$$= f (1 + \frac{v}{c} \cos \alpha) \frac{1}{1 - (v/c) \cos \alpha}$$

$$= f (1 + \frac{v}{c} \cos \alpha) (1 + \frac{v}{c} \cos \alpha + \frac{v^2}{c^2} \cos^2 \alpha + \ldots)$$

$$= f (1 + \frac{2v}{c} \cos \alpha + \frac{2v^2}{c^2} \cos^2 \alpha + \ldots)$$

Here all terms after $(2v^2/c^2)$ cos^2 α are negligibly small with regard to 1, so the received frequency can finally be written as:

$$f_v = f (1 + \frac{2v}{c} \cos \alpha + \frac{2v^2}{c^2} \cos^2 \alpha) \tag{2.3}$$

Though not indicated in *Figure 2.27*, every point of the sea-bed is hit by the beam and causes a stronger or weaker echo in the direction of the receiving transducer. All these points are situated at a different angle α to the horizontal direction; according to equation 2.3, the frequencies received aboard must differ for all these points. However, the average frequency is approximately that from point P, at an angle α to the horizontal.

Hence, *though the distance between the ship and the sea-bed does not change, the received frequency will differ (owing to the Doppler effect) from the transmitted frequency. From the Doppler frequency-shift, which can be measured, the speed v of the vessel can be found* with the aid of equation 2.3.

A second transmitting transducer directs a beam in a backward direction (*Figure 2.28*) and a second receiving transducer receives its echoes. For

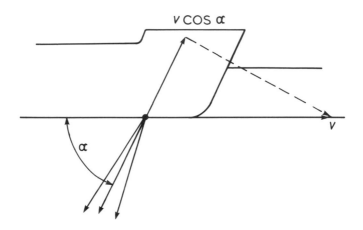

Figure 2.28 Doppler log: for the backward-directed beam the trans-ducers move away from the sea-bottom

this beam the vector $v \cos \alpha$ is negative, however, because the transducer moves away from the reflecting surface of the bottom instead of approaching it. Equation 2.3 has now changed to:

$$f_a = f \left(1 - \frac{2v}{c} \cos \alpha + \frac{2v^2}{c^2} \cos^2 \alpha \right) \qquad (2.4)$$

where f_a is the frequency received from the backward-directed beam.

The difference $f_v - f_a$ of the two received frequencies can easily be measured electronically by mixing the two frequencies and detecting the resulting beats.

From equations 2.3 and 2.4:

$$f_v - f_a = \frac{4fv}{c} \cos \alpha$$

Hence:

$$v = \frac{c}{4f \cos \alpha} (f_v - f_a) \qquad (2.5)$$

We may consider f and $\cos \alpha$ to be constant. The speed of sound waves in the water c depends, however, on the temperature and (to a smaller degree) on the salinity and the water pressure. For that reason a thermistor or velocimeter is mounted near the transducers. (A thermistor is a resistance the magnitude of which depends on the temperature.) Deviations of the sound speed c from the normal value are passed to the system computer for correction of its calculations. In equation 2.5, therefore, c, f and $\cos \alpha$ are known and $f_v - f_a$ is measured, so v can be calculated.

Note that the reading of a Doppler log depends solely on the speed of the sound waves; the propagation time of the pulse and its echo plays no role.

Automatic correction for changes in speed of sound

In some types of Doppler log (e.g. Krupp and Thomson CSF), $c/\cos \alpha$ in equation 2.5 is automatically kept constant. Krupp does so by building up each transducer from a large number (144) of electrostrictive elements of PZT material. For simplicity only four elements are shown in *Figure 2.29*.

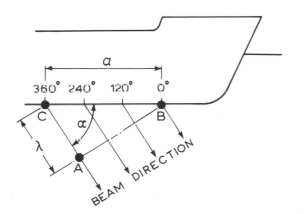

Figure 2.29 Krupp configuration of transducer elements; changes in speed of sound in water need not be corrected

If the four elements were supplied with alternating voltages in phase, the resulting sound waves would also be in phase, and the beam would be directed perpendicular to the radiating face of the transducer, i.e. vertically. However, the elements are fed with voltages that differ in phase by $120°$, so the sound waves have the same phase difference. At all points of the line AB, however, the sound vibrations are in phase. Such a line or plane is called a wave front; propagation is always perpendicular to a wave front. Hence, cos $\alpha = \lambda/a$ (*a* being the distance between the first and last elements). As $\lambda = c/f$ it follows that

$$\cos \alpha = c/af \text{ and } c/\cos \alpha = af \tag{2.6}$$

Because *f* and *a* are constant, $c/\cos \alpha$ must also be constant. If we substitute *af* for $c/\cos \alpha$ we find:

$$v = \frac{af}{4f}(f_v - f_a) = k(f_v - f_a)$$

where $k = a/4$. From this formula it follows that *v* is proportional to $f_v - f_a$. The thermistor for measuring *c* is now superfluous.

Reflections

Both the echo sounder and the Doppler log react to reflections of sound waves from the sea-bed; the former measures the propagating time and the latter the difference of the two frequencies $f_v - f_a$.

If the beam is propagated from one water layer into a second one of different composition or temperature, there will be reflection; there will also be a Doppler effect if the second layer moves relative to the first layer and if the beam hits this layer obliquely. In that case the frequency of the sound vibrations penetrating the second layer will also change, if the speed of the sound waves in the second layer is different from that in the first layer ($f = c/\lambda$). For the echo, however, the reverse frequency change will occur and will cancel out the first change.

A Doppler log measures the algebraic sum of all Doppler frequency shifts experienced by the sound on its way to the bottom (or to a reflecting layer) and back again. To this frequency shift must be added the shift that arises at the transition of the transducer vibrations between the ship and the water, and vice versa.

If the beam hits the bottom (bottom contact) the total frequency shift is, according to equation 2.5, proportional to the speed of the ship with regard to the bottom. If there is no bottom contact, but only reflection against a water layer, the measured Doppler shift is proportional to the speed of the ship relative to that water layer.

Janus configuration

The placing of the two transmitting transducers, as described above, to produce forward and backward beams is called a Janus configuration. (This name refers to the way the transducers look forward and backward, like the god Janus of the ancient Romans, who was represented with two faces in order to show that he looked into the past and into the future.) Thanks to the Janus configuration a linear relationship exists between the speed of the vessel v and the measured frequency shift $f_v - f_a$. A further advantage is that vertical movements of the ship cause equal changes to the Doppler shifts f_v and f_a in the forward and backward beams, so the difference $f_v - f_a$ remains the same. *Vertical movements of the ship do not therefore influence the Doppler shift.*

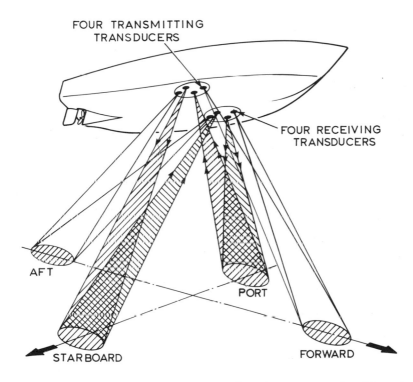

Figure 2.30 Pairs of transducers in Janus configuration to measure alongships and thwartships speeds

For measuring the thwartships speed, a similar Janus configuration is mounted at an angle of $90°$ with the alongships transducers; see *Figure 2.30*. The distance from the bridge of a large tanker to the bows may be 250 metres, so special information about the thwartships speed both fore and aft is required when mooring. In that case thwartships transmitting and receiving transducers are mounted both fore and aft, as shown in *Figure 2.31*.

Pitching and rolling

In *Figure 2.32* the dotted lines represent a ship's possible change of position due to pitching. (It is true that a ship does not 'rotate' around point
P, but because a Janus configuration does not react to vertical movement
we may take it as such.) From the figure we see that the speed v_1 =
$v \cos \alpha$ for the dotted position of the ship, and for the forward-directed
beam v_1 increases to v_1'; for the backward-directed beam v_1 decreases to
v_1''. If the ship 'rotates' in the opposite direction, v_1' will be smaller than
v_1 and v_1'' will be larger.

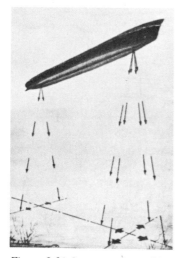

Figure 2.31 In a very large ship,
thwartships speed needs to be
measured both fore and aft

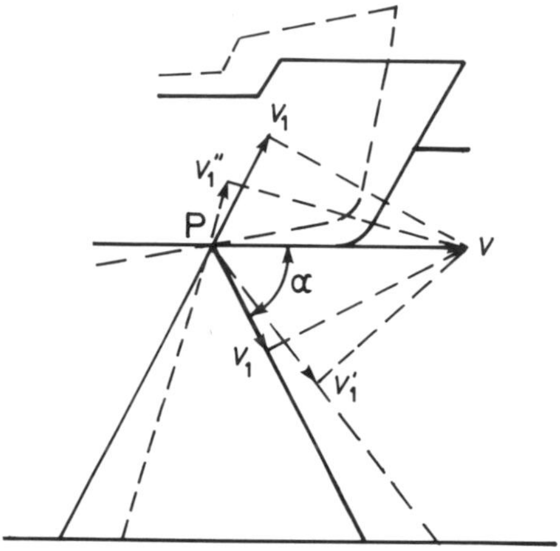

Figure 2.32 During pitching, v_1' increases and
v_1'' decreases (or vice versa); Doppler shift is
not affected, because $v_1' + v_1'' = 2v_1$

Equation 2.5 is obtained by taking the difference of the Doppler shifts
$+2fv_1/c$ for the forward beam and $-2fv_1/c$ for the backward beam. In the
horizontal position of the ship this difference is $4fv_1/c$, and in the turned
position it is $2fv_1'/c + 2fv_1''/c = (v_1' + v_1'')2f/c$. When v_1' becomes
smaller v_1'' becomes greater, or vice versa, so $v_1' + v_1''$ is approximately
$2v_1$. Hence *the Doppler measurement of the speed is not, in practice,
influenced by pitching*. The same applies to the two thwartships beams
during rolling.

Continuous-wave and pulse systems

Hitherto it has been taken for granted that the transmitting transducers
generate vibrations continuously, thus making it necessary for each beam

to have a separate transmitting and receiving transducer. This is called a *continuous-wave (c.w.) system*. Transmitting and receiving transducers are of identical construction.

Other types are *pulse systems*. In such a system a transducer generates pulses and the *same* transducer receives the echo between the transmissions. Therefore a pulse system needs only half as many transducers as a c.w. system.

With c.w. systems the reception of the echo can be disturbed by the continuously emitted vibrations of the transmitting transducer going directly from transmitting to receiving transducer (so-called cross-noise or feedback). With pulse systems this cannot occur, since a pulse is transmitted only after the echo of the preceding pulse has been received, and the receiver is blocked during the transmission.

Further advantages and disadvantages of the two systems will not be dealt with here. The majority of Doppler logs in use are pulse systems.

Transducers

The angle of the alongships beams is about 3°, that of the thwartships beams about 8°. The frequency used is 100 to 600 kHz. The surface area of each transducer need then be only about 10 cm². The high frequency

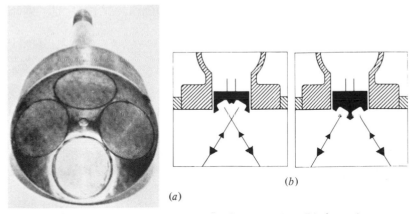

(b)

(a)

Figure 2.33 (a) Four transducers in integrated construction; (b) alternative arrangements of Janus-configuration transducers

and the concave shape of the surface also lead to a small beam angle. The higher frequency influences the reflection and the absorption but not the speed of propagation.

The transducers are of the electrostrictive type. *Figure 2.33(a)* shows an example of four transducers in a combined construction, and *Figure 2.33(b)* two possibilities for a Janus configuration. Usually the transducers are inserted in a 'sea chest' or 'sea well' (*Figure 2.34*), permitting their removal for repairs or replacement without the ship requiring dry-docking. The diameter of the hole required in the hull plates is about 350 mm.

Replacement of a transducer (1) in a sea chest without the ship being dry-docked can be done in the following way. After the transducer (which is connected to the other apparatus by means of a cable with a plug and socket) has been disconnected, some nuts (2) are loosened and the bolts turned in the direction of the arrows. Now the transducer (1) can be drawn upwards until it is above the flange (4) in the upper part (3) of the sea chest. This upper part is then shut off from the lower part (6) by

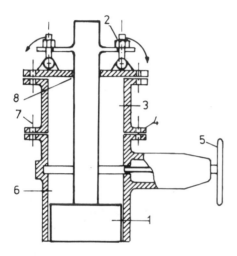

Figure 2.34 Sea chest containing a transducer

means of a sliding valve operated by the handwheel (5). In order to check that the valve is properly shut a tap (not shown in the figure), connected to the upper part (3), can be opened. If the water in the upper part is not under pressure the bolts (7) of the flange (4) may be removed. By using grease, the transducer can be slid easily from the top flange (8). The sequence is reversed when a new transducer is mounted.

Measurement of ship's speed relative to bottom or to water

Owing to absorption by particles in the water at a depth of 200 to 400 metres, a Doppler log only functions down to about 200 metres. When sufficiently low frequencies are used, echoes may still arrive from a rocky bottom at a depth of 600 metres and more. In general, however, the beam is absorbed and scattered by the mass of water between 200 and 400 metres, the so-called *deep scattering layer* (DSL). When reflections are received from this layer the speed of the ship relative to that layer, and not relative to the bottom, is obtained. Thus uncertainty and confusion may occur.

Apart from the effect of the DSL, the water at 10 to 30 metres below the keel also causes an echo and Doppler effect by volume-reverberation; this is called 'water track' (as opposed to 'bottom track'). In deep water there is a considerable difference between the time of propagation for bottom reflection and that for reflection from the mass of water at a depth of 10 to 30 metres. Receivers can be made operative for only a short period (a certain 'window' of time) either immediately after or a short time after each pulse transmission. Suppose that the receiver has bottom contact, with the window occurring a short time after transmission. If the Doppler log then loses bottom contact, the window is automatically shifted to occur immediately after pulse transmission. As a result, the receiver reacts only to reflections from the 10–30 metre water layer. When this happens, the light in the 'bottom track' pushbutton is extinguished and the 'water track' pushbutton lights (see *Figure 2.36*). In the Krupp Atlas Doppler log, for depths less than 600 metres it is possible to switch manually to the water track mode.

Uses of the Doppler log

The Doppler log has received acceptance in the navigation of very large tankers. These vessels need navigational accuracy in coastal waters that cannot be completely guaranteed by shore-referred navigational aids. A second very important advantage in large tankers is the use of the log for berthing or mooring to jetties, single-point mooring, low-speed cruising, dropping or weighing anchor, and approaching fixed docks. The size of such vessels has increased so much that the reliability of human judgment concerning speed and distance is considerably reduced. There have been numerous docking accidents due to incorrect estimates of ship's approach speed. For example, for a tanker of 200 000 tonnes with a residual speed when tying up of 0.2 knots (0.1 m/s), the energy to be absorbed by a pier or dolphin together with the ship's side is $\frac{1}{2}mv^2 = 1\,000\,000$ joules. The Doppler log can measure the speed to the nearest 0.01 knot or 5 mm/s; unfortunately, however, it sometimes does not function correctly during docking if the screws of tugs cause air bubbles (which reflect sound waves) to pass through the beams. In docking it is desirable to measure the thwartships velocity separately fore and aft, so transducers are placed fore and aft. To allow for the position of the master during docking, two large-digit displays of velocities are mounted on the bridge wings, with the control unit on the bridge itself (*Figures 2.35* and *2.36*). In order to avoid loss of anchor on large vessels, the speed relative to the bottom during anchoring must be extremely low, and can be determined accurately by the Doppler log.

Figure 2.35 Doppler-log display unit showing thwartships speed (knots), alongships speed (knots) and distance travelled (kilometres) (courtesy Krupp Atlas)

Figure 2.36 Doppler-log control unit, including controls for bridge-wing display units (courtesy Krupp Atlas)

In certain harbour approaches the speed is subject to an officially prescribed maximum. The Doppler log can measure this speed accurately. A further application is for geophysical surveys in connection with exploration for minerals, if the area concerned is not within the coverage of a Decca Hi-Fix or a Decca chain. The Doppler log is then used in conjunction with the satellite navigation system. The Doppler log provides the speed input for the satellite receiver and for dead-reckoning tracking between satellite fixes (it normally produces 200 pulses per nautical mile). It also determines drift from anchor position during non-operational periods.

Technical data

The technical performance of a Doppler log depends, of course, on the make. The following list of average performances is based on data supplied by Ametek, Edo-Western, Furuno, Krupp, Magnavox, Sperry, True Digital and Thomson CSF.

Accuracy: 0.2 to 0.5 per cent of the distance travelled, plus 20 metre/hour drift of the set. For high speeds, e.g. 20 to 40 knots, about 1.0 per cent.

Velocity range: up to between 30 and 100 knots alongships, up to between 8 and 10 knots athwartships.

Minimum depth: about 0.3–0.5 metre.

Frequency of vibrations in water: 100 to 600 kHz.

In the future the use of the Doppler log will probably not be restricted to very large ships. This is because the log is very accurate and is independent of weather conditions, current (provided there is bottom contact), depth of the water (up to a certain limit), pitching, rolling (up to 8°), trim (up to about 3°), velocity of the ship and temperature of the water. The limit to further increase in accuracy of dead-reckoning fixes (as desired, for instance for geophysical surveys) is not the Doppler log itself but the additional necessity of a gyro compass.

After a gale the water can be saturated with air for many days, especially in wintertime. If the ship does not move relative to the water, its heat is transferred to the water near the hull; thus the temperature of the water is increased somewhat. At the higher temperature the water loses air and bubbles form, preventing the Doppler log from functioning correctly. In particular, the drift shown on the indicator will increase considerably. The same phenomenon can happen if the ship is in port and gas (methane) rises from the bottom; this effect does not occur when the ship moves, even at a very slow speed, or when the ship is anchored but the water is moving. There is also the problem (mentioned on page 73) of air bubbles below the keel when going astern.

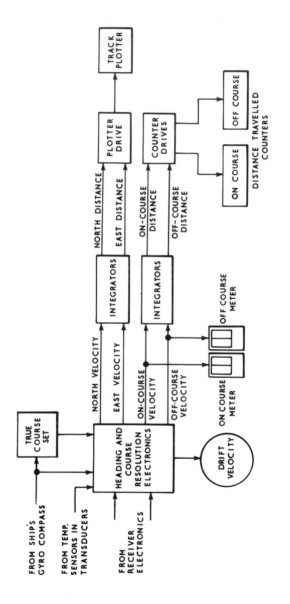

Figure 2.37 Block diagram of Sperry Doppler log system

Block diagram

When the Doppler log is used for navigation, the alongships and thwart-ships velocities are supplied to a computer; see *Figure 2.37*. The gyro compass is also connected to the computer. Unless the magnitude of $c/\cos \alpha$ is automatically kept constant, the water temperature is supplied to the computer.

The computer in turn is connected directly to two indicators for the 'on course' and 'off course' velocities and one for the drift. The computer also calculates the north-south and east-west components of the velocity, derived from Doppler log and gyro compass, and supplies them to an integrator, which calculates the distances covered in the N-S and E-W directions; these distances are then supplied to a track plotter, via a 'plotter drive'. The 'on course' and 'off course' velocities are integrated to obtain the distances involved.

THE ELECTROMAGNETIC LOG

The electromagnetic log is based upon the Faraday-Maxwell induction law; *Figure 2.38* shows the principle of the log. A direct current through the windings of a coil generates a magnetic field. Four conductors (ab, bc, cd

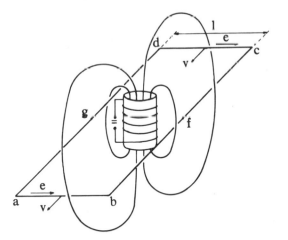

Figure 2.38 Electromagnetic log principle: when loop abcd moves at velocity v, a direct voltage is induced in ab and cd, proportional to the velocity

and da) are arranged in the form of a loop around the coil. If the conductors do not move relative to the coil they do not intersect the magnetic lines of force and no voltage is induced in them.

Suppose that we move the loop in the direction of the arrows at a velocity v. (We will assume that the distance of the conductor cd from the coil is sufficiently great that it can be moved at this velocity for some time without touching the coil; also that the intensity of the magnetic field is the same everywhere in the plane formed by the four conductors, and that the lines of force are perpendicular to the conductors.) As a result of the velocity, there arises in the conductors ab and cd (and not in bc and ad) a voltage $e = Blv$, where B is the magnetic induction and l is the length of ab and cd. Since B is practically equal to the magnetic field strength H, $e = Hlv$.

The direction of the voltage e depends on the directions of the lines of force and the direction of the velocity of the conductors ab and cd. Both voltages e are equal and have the same direction but neutralize each other in the loop, so there is no current. If, for instance, e is 1 volt in each of the two conductors, a voltmeter between any point of ad and any point of bc will indicate 1 volt.

According to the formula the induced voltage is proportional to the velocity v. Should the velocity have the opposite direction, the direction of the voltage would change too.

Alternating current through the coil

Instead of a direct current, suppose that we send an *alternating* current through the coil. The field intensity H now changes to $H_m \sin \omega t$ and the induced voltage becomes $e = H_m lv \sin \omega t$, or $e = E_m \sin \omega t$ if we write E_m instead of $H_m lv$. So we have an alternating voltage with an amplitude E_m that is proportional to the velocity v. For the electromagnetic log an alternating voltage is preferred to a direct voltage.

Application of principle in electromagnetic log

The coil of an electromagnetic log is inserted in a watertight 'flow probe', 'rod' or 'flow sensor' (*Figure 2.39*), which projects through the hull into the water. The construction is usually such that, in order to avoid damage when navigating in shallow waters, the sensor can be retracted into a tube by electrical control from the bridge.

The flow sensor has a streamlined shape. This not only decreases the resistance experienced by the sensor on a moving vessel, but, more important, it prevents as much as possible the surrounding water from being dragged with the ship. Since the ship's speed is measured relative to the water surrounding the sensor, dragging the water would influence the accuracy of the indicated speed.

The conductors forming the loop in *Figure 2.38* are in reality the surrounding water, which behaves as a conductor. The water moves, relative to the coil, at the ship's velocity v. The water flowing away makes room for other water; this means that, in electrical terms, in spite of the water flowing away the conductors remain in the same spot.

On both sides of the sensor an electrode is fitted for electrical contact with the water (points g and f in *Figure 2.38*); one of these contacts can be seen in *Figure 2.39*. Internally the contacts are connected to the

Figure 2.39 Sensor of electromagnetic log (courtesy Sperry Marine Systems)

apparatus for measuring the induced voltage. Note that the contacts should be cleaned about once a month.

Figure 2.40 shows that the speed indicator is connected via a junction box to the master unit, where the signal from the flow sensor is amplified, and can also be corrected; the magnitude of the correction is ascertained by calibration trials. The velocity can be converted in the master unit from analogue to digital form, if the readout is a digital one. It is possible to connect more speed indicators into the system, for mounting at various positions aboard.

The speed-indicator signal can be supplied to a special generator, which produces a number of pulses per minute proportional to the magnitude of the signal and hence to the distance travelled. It amounts to 200 pulses for

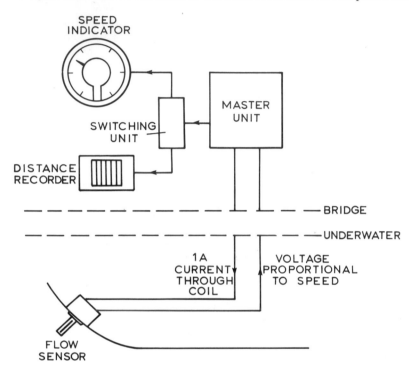

Figure 2.40 Block diagram of electromagnetic log system

each nautical mile. The generator is connected to a 'distance run' recorder with six digits, resembling the mileage-indicator of a car.

Depending on the make, the range of the speed indicator is 0 to 25 (even up to 70) knots forward, sometimes with a much smaller range for speed astern. The accuracy of the indicated distance is about 1 to 2 per cent.

3

Direction finders

A direction finder consists of a receiver and aerial system used together to determine the direction of incoming radio waves. On board a ship, if the angle between this direction and a known direction (e.g. true North) is found, and if the position of the transmitting station is known, it is possible to draw on the chart a line on which the ship is situated. By taking a bearing from the second station, a second position line is found. The point of intersection of the two lines can be assumed to be the position of the ship.

With few exceptions, direction finders are obligatory for all ships of 1600 tons gross tonnage and over. However, smaller vessels too are nearly always equipped with them.

Some advantages of a direction finder as compared with other position-fixing systems are as follows:
1. Bearings can be taken of *any* transmitting station.
2. The direction finder can be used for coastal navigation anywhere in the world.
3. Ships in distress that can still make use of their radio transmitters can have their bearings taken directly by other ships and therefore can easily be found.

In many places on shore near important shipping routes, and on most lightvessels, special transmitters have been erected to enable ships to take radio bearings. These are the so-called radio beacons, which operate in the special radio-beacon band of 285—315 kHz.

Direction-finding stations have also been erected on shore to provide bearings for ships that have no direction finder but are equipped with a wireless transmitter and receiver. A drawback of this system is that in foggy weather, when bearings are most needed, the ships must sometimes wait a long time before it is their turn to have their bearings taken. As practically all ships now have direction finders, the number of direction-finding stations has considerably decreased. In 1977 there were 12 left, as against about 840 radio beacons.

PRINCIPLE

It is well known that if a changing number of magnetic lines of force pass through a coil a voltage is induced. It is immaterial where these lines of force have originated. The alternating current in a transmitting aerial also sets up lines of force, which are propagated at the velocity of light. In order to make use of this radiation to take a bearing from a transmitter, a loop aerial is used instead of a coil. *Figure 3.1* shows a loop aerial that

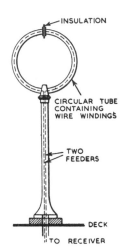

Figure 3.1 Loop aerial rotatable on vertical axis (only one winding is indicated)

can turn on a vertical axis. Inside the circular tube are a number of large windings, so that this aerial can be considered as a large coil. It should be noted that the dimensions of the coil are small in comparison with the wavelength, which for radio bearings is about 1000 metres.

When such an aerial is in the field of radiation of a transmitter (see *Figure 3.2*), magnetic lines of force pass through the windings. As the number of these lines alternately increases and decreases, an alternating voltage is set up in the coil. The voltage set up in the loop is greatest in the position shown, as the plane of the loop is turned towards the transmitter so that a maximum number of lines of force pass through the windings. If the loop aerial is joined to a receiver by the connections shown as broken lines in *Figure 3.1*, the strongest signals from the transmitter will be heard with the aerial in the position shown in *Figure 3.2*.

When the loop is rotated about a vertical axis through a quarter of a revolution, no lines of force will pass through the coil, so no alternating voltage is set up and the transmitter is no longer heard. When the loop is rotated through a further 90°, a maximum number of lines of force will again pass through it and the signals received will return to a maximum.

During one revolution of the loop aerial the signals will, therefore, twice attain a maximum strength and twice vanish.

We shall now examine the reception in any position of the loop. *Figure 3.3* is a top view of a loop aerial that can rotate about the vertical axis O. (The circle does not represent the loop.) The plane of the loop aerial is in an arbitrarily chosen position BC with respect to the transmitting aerial, which is supposed to be situated at a great distance to the left. The

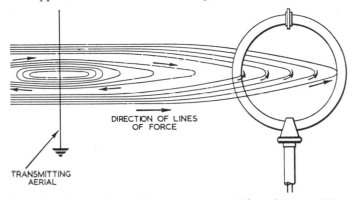

Figure 3.2 Magnetic lines of force are propagated from the transmitting aerial; the number of lines of force that pass through the loop aerial varies, and an alternating voltage is set up in the windings inside the tube

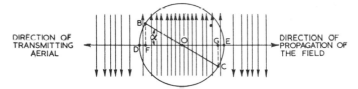

Figure 3.3 Plan view of loop aerial. With the loop (rotatable about O) in the position BC, the number of lines of force passing through it is cos α times as many as in the position DE; the voltage induced is therefore cos α times as great

distance to the transmitter being large compared with the dimensions of the loop aerial, the magnetic lines of force shown in the figure may be considered as vertical straight lines. The number of lines enclosed by the loop in the position BC is smaller than in the position DE. The ratio of these numbers is:

$$\frac{FG}{DE} = \frac{FO}{DO} = \frac{FO}{BO} = \cos \alpha$$

Consequently the voltage set up in the position BC is also $\cos \alpha$ times the voltage E_{max} in the position DE, where E_{max} is the maximum alternating voltage.

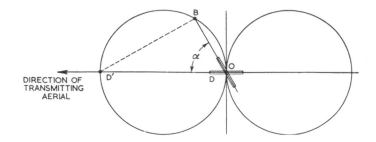

Figure 3.4 Figure-of-eight polar diagram of the voltage set up in the loop by the transmitter. When the loop is in the position OD, with its plane in the direction of the transmitter, the voltage is represented by OD'. When the loop is rotated through angle α, the voltage becomes
$$OD' \cos \alpha = OB$$

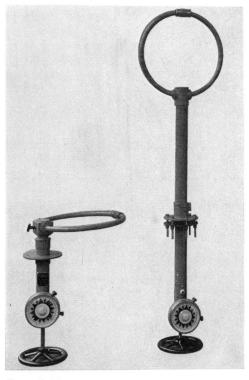

Figure 3.5 Loop aerials rotated by handwheels; the bearing is read off the scale (courtesy Radio Holland)

In *Figure 3.4* the alternating voltage for various positions of the loop is represented by a polar diagram. The transmitter is assumed to be in the direction indicated. When the loop is in the position OD, the voltage set up in it, represented by OD', is greatest. If the loop is rotated through an angle α, the voltage is, according to the above, reduced to OB = OD' cos α. The point B must be situated on a circle the diameter of which is OD', for in that case the angle OBD' is a right angle and OB = OD' cos α. The polar diagram showing the alternating voltage set up in the loop aerial for any position of the latter consists, therefore, of two circles and is called a figure-of-eight diagram.

To determine the direction of a transmitter, the loop can be rotated to the position for maximum signal strength. The transmitter then lies in the extension of the plane of the loop. It is also possible to rotate the loop so that the transmitter is no longer heard, in which case its direction is at right angles to the plane of the loop. In practice, the loop is always rotated so that no signals are heard. This is done because a small change in the position of the loop will result in rapidly increasing signal strength and so the exact place of the minimum will be easy to determine. Thus the minimum is sharp. If, on the contrary, the position for maximum signal strength were used, a small change would cause only a very slight decrease in signal strength, so that the exact place of the maximum cannot be readily fixed. This is evident from *Figure 3.4*.

The axis of the loop passes through the deck to the chart-room or the wireless cabin, so that the loop can be rotated from one of these places. A pointer is attached to the axis and moves over a circular scale. Sometimes the rotation of the loop is transmitted to the pointer through a gearing (see *Figure 3.5*). Alternatively, the scale may be attached to the axis and the pointer fixed. The scale is graduated in a clockwise direction from $0°$ to $360°$. A second scale with a division into the points of the compass may also be supplied.

Sense determination

When a transmitter is, for instance, straight ahead and the loop is rotated so that no signals are heard, the pointer indicates zero on the scale (or $180°$, if the loop has been rotated to the other minimum). The direction of the transmitter is then perpendicular to the loop, in one of two alternative directions, $180°$ apart. The ambiguity is resolved (*sense determination*) in the following way.

A simple aerial, consisting of a vertical wire, receives equally well from all directions, so its polar diagram is a circle (*Figure 3.6*). For sense determination such a vertical or auxiliary aerial is used in addition to the loop aerial with its figure-of-eight diagram. Care must be taken that reception via the vertical aerial is equal to reception via the loop when the latter is in the position for maximum signal strength. The diameter of the small circles in *Figure 3.6* will then be equal to the radius of the large circle. Simultaneous reception via both aerials thus gives a diagram that is the sum of a circle and a figure-of-eight.

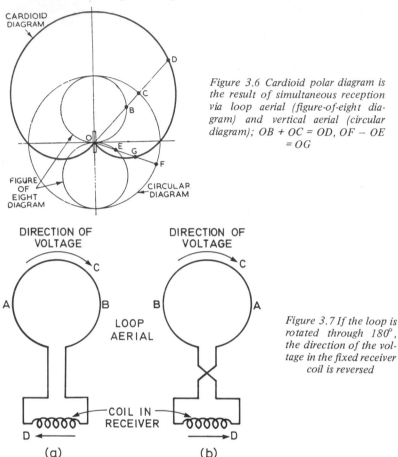

Figure 3.6 Cardioid polar diagram is the result of simultaneous reception via loop aerial (figure-of-eight diagram) and vertical aerial (circular diagram); OB + OC = OD, OF − OE = OG

Figure 3.7 If the loop is rotated through 180°, the direction of the voltage in the fixed receiver coil is reversed

It should be noted that, although the alternating voltage set up in the loop in any position is equal to that induced in it in a position half a rotation away, there is a 180° phase difference between the two voltages. This is shown by *Figure 3.7*. If the loop of (*a*) is rotated through 180°, (*b*) is obtained. The direction of the voltage set up in the loop at a given moment is indicated by the arrows C. It will be seen that the voltage applied to the coil in the receiver (arrows D) is opposite, at that same moment, with the loop rotated; this means that the phase of the voltage supplied to the receiver changes by 180°. The alternating voltage set up in the loop

(*Figure 3.6*), represented by the upper circle of the figure-of-eight diagram, may be *added* to the alternating voltage of the vertical aerial, which is in phase with it, and which is represented by the circle diagram (OB + OC = OD). The voltage represented by the lower circle of the figure-of-eight diagram (being 180° out of phase with that of the upper circle) should be *subtracted* from the voltage set up in the vertical aerial (OF − OE = OG). The diagram obtained in this way, shown in *Figure 3.6* with a thick line, is called the *cardioid* (heart-shaped) diagram.

To obtain the signals from both the loop and vertical aerials (cardioid diagram), a switch with the two positions 'sense' and 'direction finding' (or 'DF') must be set to 'sense'. The signals received via the vertical aerial are induced in coil S by coil L (*Figure 3.8*). The resistance R in the vertical

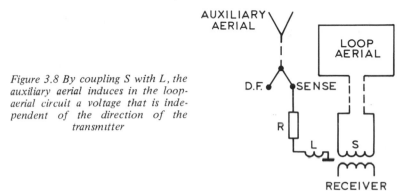

Figure 3.8 By coupling S with L, the auxiliary aerial induces in the loop-aerial circuit a voltage that is independent of the direction of the transmitter

aerial reduces the reception in order to make the radius of the circle diagram equal to the diameter of each of the figure-of-eight circles.

The cardioid has only one minimum and one maximum. Hence it can be used to remove any doubt about the direction of a transmitter. There are various ways of doing this.

In one type of direction finder, a second pointer (the sense pointer) is fitted at an angle of 90° to the first pointer (the d.f. pointer); the sense pointer is used for the cardioid diagram and the d.f. pointer for the figure-of-eight diagram. With the switch set to 'sense' and the loop rotated to the minimum position of the cardioid, *the sense pointer indicates on the scale the angle between the ship's head and the direction of the transmitter*. The minimum of the cardioid is, however, less sharp than that of the figure-of-eight diagram, and the cardioid may also be distorted from the ideal shape shown in *Figure 3.6*. Hence *the cardioid must not be used for taking bearings* but only for sense determination. To prevent the taking of bearings the sense pointer is purposely given such a form that it cannot be used to read the scale accurately.

The minimum of the cardioid diagram having been found, the switch is set to the position 'DF' so that signals are obtained only from the loop aerial (figure-of-eight diagram). The signal strength will increase suddenly because the minimum of the cardioid coincides with one of the maxima of the figure-of-eight diagram. The loop now has to be rotated 90° along the shortest way to the position of minimum signal strength. The ambiguity has, however, been resolved because the sense pointer has indicated the correct sense. *So the d.f. pointer should be turned to the position first occupied by the sense pointer.* The position at which no signals are perceptible is determined as accurately as possible; apart from corrections, the reading is the bearing of the transmitter with respect to the ship's head.

In other types, the sense is determined in the following way. The loop is first rotated to one of the two minima of the figure-of-eight diagram; one of these two minima (always the same) indicates the correct bearing, for instance that to the right of O in *Figure 3.6*. Then the auxiliary aerial is switched on ('sense' position) so that the cardioid diagram applies; there will be a sudden increase in sound volume, caused by reception solely via the auxiliary aerial. The loop should then be rotated in a clockwise direction (i.e. to higher numbers on the scale). The sound will now decrease, proof that the loop was initially set at the *correct* minimum; the pointer then indicates the correct transmitter bearing. If the sound *increases* the loop was set at the *wrong* minimum.

It should be noted that it is not usually necessary to determine the sense of the bearing, for as a rule, when bearings are taken, there is no doubt about the sense. It may be necessary, however, when taking a bearing from a ship in distress or when approaching a lightvessel.

The Bellini-Tosi system

The rotating-loop direction finder has the drawback that the loop must be installed exactly above the place from which it is operated, i.e. the wireless cabin or the chart room, and this is not only sometimes impossible but in most cases also not the best place from the point of view of the accuracy of the bearings. In the Bellini-Tosi (B-T) system this drawback is largely overcome.

Figure 3.9(a) shows two fixed frames fitted at right-angles to each other. The direction of the plane of one of the frames is alongships, that of the plane of the other is thwartships. The beginning and end of the windings of each frame are connected to separate coils in the receiver (the alongships and the thwartships field coils, which are similarly at right-angles), so that four conductors lead from the B-T frame to the receiver. In *Figure 3.9(b)*, for clarity, only one winding of each frame and field coil has been drawn.

Suppose that there is a transmitting station on the starboard side at a bearing of 45° to the ship's head (*Figure 3.10*). The alternating voltages set up in the two frames are then equal and, since the frame circuits are identical, the currents in them are equally strong; they will also be in phase.

An alternating current in a coil produces an alternating magnetic field proportional to the current. The field may be represented by a vector with a length proportional to the strength of the field (and thus of the current) and with a direction the same as that of the field. Within the field coils two such fields are therefore produced, at right-angles to each other and, in the case under consideration, equally strong. These fields combine into

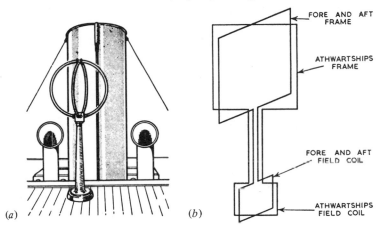

Figure 3.9 Bellini-Tosi aerial: (a) typical installation; (b) schematic showing two fixed frame aerials and two field coils

one resultant field (*Figure 3.10*), *the direction of which is the same as that of the lines of force caused by the transmitter in the B-T aerial.* Calculation shows that this is always the case, whatever the direction of the transmitter. Note, in *Figure 3.10*, that the field coil connected to the thwartships frame is fitted in the alongships direction, and the other field coil in the thwartships direction.

Inside the two field coils we may now place a small rotatable coil, called the search coil (see *Figure 3.11*). The whole is termed a *radiogoniometer*. If we turn the search coil so that most of the lines of force of the resulting alternating field pass through it, the alternating voltage set up in it by these lines of force is a maximum. As the ends of the search coil are connected to the receiver via a pair of sliprings and the contacts resting on them, the signals are then strongest. If the search coil is then rotated through 90°, no lines of force will pass through it and no signals will be heard.

Just as the intensity of the alternating voltage set up in a rotatable frame aerial can be represented by a figure-of-eight diagram, the alternating voltage set up in the search coil can also be represented by a figure-of-eight diagram.

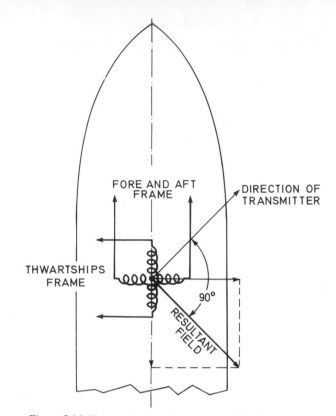

Figure 3.10 The production of a resultant field from the two fields set up in the receiver by a transmitter at a bearing of 45° from ship's head

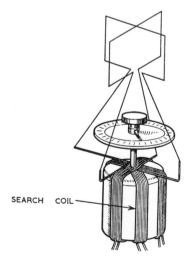

Figure 3.11 A rotatable search coil, fitted inside the two field coils, determines the direction of the resultant field of Figure 3.10

THE PRINCIPAL ERRORS AND THEIR CAUSES

Inaccuracies in the bearings arise from various causes, described below.

Errors caused by surroundings

It stands to reason that the transmitter from which we intend to take a bearing induces an alternating voltage not only in the frame aerial, but also in the hull of the ship, the masts, the funnel, the rigging, the stays, the derricks, etc. Consequently, in all these interfering conductors, small alternating currents appear which, in their turn, give rise to radio waves. (This may also be called 'reflection'.)

All sorts of interfering fields will arrive at the frame aerial in this way. One consequence of interfering fields is that the sound is not zero in any position of the frame or the search coil. The exact point of the minimum cannot therefore be easily determined, or in other words the minimum is 'dim'. Another consequence is that the direction of the minimum is displaced. We can obtain a sharp minimum again by exact adjustment of a special 'zero cleaning' or 'zero sharpening' control.

The procedure for taking a bearing is as follows: first make sure that the knob for zero sharpening is in the centre position (at $0°$ of its scale), and turn the frame or the search coil till the signals are weakest. Then the signals are made still weaker with the aid of the zero sharpening control, after which the frame or search coil and then again the zero sharpening is adjusted, and so on. Zero sharpening and frame (or search coil) are finally in the exact position if the signal grows stronger at the least rotation of the frame (or the search coil) in either direction. Yawing of the ship causes the bearing to change with respect to the ship's head, so the manipulations should be done quickly.

The interfering fields are caused mainly by currents in the ship's hull. The lines of force set up by these currents 'draw' the bearing more or less towards the fore-and-aft axis of the ship, and compensating corrections must be applied. These are greatest for bearings of $45°$, $135°$, $225°$ and $315°$, and zero for $0°$, $90°$, $180°$ and $270°$ (*Figure 3.12*). For example, if the bearing of a station according to the reading of the direction finder is $40°$, when its real bearing is $45°$, the correction is $+5°$. (A positive correction must be *added* to and a negative correction *subtracted* from the reading.)

As the error is due to a deviation towards the fore-and-aft direction, the correction is positive in the first quadrant ($0°-90°$), negative in the second ($90°-180°$), positive in the third ($180°-270°$), and negative again in the fourth ($270°-360°$). The corrections change sign four times from $0°$ to $360°$ and are known as *quadrantal corrections*. *Figure 3.13* is an example of a quadrantal correction curve.

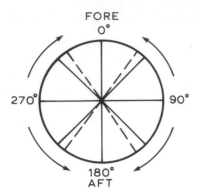

Figure 3.12 Dashed lines indicate bearing readings, which deviate towards the fore-and-aft axis; solid lines indicate real bearings after correction

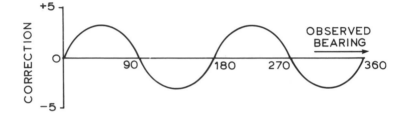

Figure 3.13 Quadrantal correction curve

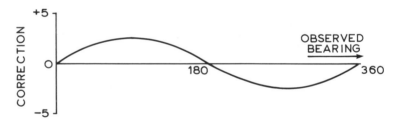

Figure 3.14 Semicircular correction curve

Some interfering conductors (for instance, an aerial tuned to a broadcasting station) can produce a very strong interfering field, particularly if the frequency of the broadcasting station is approximately equal to that of the transmitter whose bearing is required. The correction to compensate for this changes sign twice over 360°. It is therefore called *semicircular correction. Figure 3.14* shows a semicircular correction curve.

By taking radio bearings and optical bearings from transmitting stations in different directions and comparing them, the *total correction* (quadrantal, semicircular and other corrections) can be determined and represented by a curve. As the quadrantal error may be as much as 18°, it is clear that each ship needs a correction curve, to be applied to any bearing before it is made use of for navigation.

In the B-T system the quadrantal error is compensated entirely or partly by connecting a 'calibrating coil' in parallel across the alongships field coil. Expert personnel can adjust the number of windings so that the quadrantal error becomes zero or at least much smaller. There may even be too much compensation, resulting in errors of the opposite sign. *Adjustment of the calibrating coil should be made only by expert personnel.*

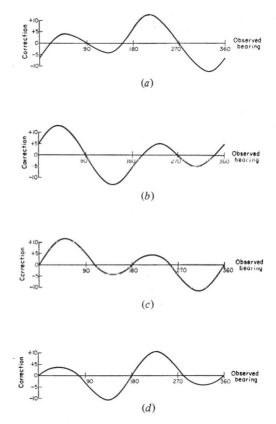

Figure 3.15 Combined quadrantal and semicircular correction curves, the semicircular component being caused by a resonance aerial or other interfering conductor near the d.f. aerial: (a) on its starboard side, (b) on its port side, (c) behind it, (d) in front of it

Obviously, a new correction curve has to be plotted when the windings of the coil are altered, and this entails much work.

Compensation for quadrantal errors can also be made in rotatable-loop direction finders, but this is not usually done.

The occurrence of errors is not serious, provided that their extent is exactly known and a correction applied. The corrections depend on the interfering conductors. Conversely, the shape of the correction curve can indicate the position of interfering conductors. This is shown by *Figure 3.15(a) to (d)*.

Night effect

At frequencies normally used for taking radio bearings, sky-wave reception predominates at night at longer distances from the transmitter. During the day the ionosphere is ionized more intensely by sunlight, and the loss suffered in it by the usual radio-beacon frequencies is much greater than that by night, with the result that, by day, at distances as far as about 200 miles, only the ground wave is perceptible.

The presence of sky waves at night has two unfavourable effects. In the first place, it causes fading (see page 40), which is caused by simultaneous reception of both ground and sky waves, and makes the adjustment at minimum sound strength more difficult. Secondly (and this is more serious) the reception of sky waves results in bearing errors; these can theoretically be as high as 90°, but are generally limited to 4° or less.

Night effect may appear from one hour before sunset till one hour after sunrise — and, according to observations, chiefly during dusk — if the distance to the transmitter exceeds about 25 miles. If less accuracy is allowed, it may be assumed that up to 100 miles no night effect will appear. Note that, though this phenomenon is called night effect, it may also appear during the day at a distance from the transmitter that exceeds about 200 miles; in winter the ionization of the ionosphere at latitudes higher than 65° is so weak, with the exception of some hours in the middle of the day, that night conditions always prevail. Also, in very rare cases, night effect may appear when there are high mountains between the transmitter and the direction finder.

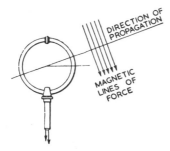

Figure 3.16 Rotation of lines of force as a result of ionospheric reflection

The cause of the night effect is as follows: when the waves are bent downwards in the ionosphere (chiefly in the E-layer), the magnetic lines of force may be turned from the horizontal (while remaining perpendicular to the direction of propagation). Suppose, for instance, that ionospheric reflection turned the lines of force through 90°, as shown in *Figure 3.16* (i.e. the lines of force are in the same plane as the paper, instead of being perpendicular to it). If the loop aerial were then rotated to the position in which reception is zero, from that position we would conclude that the transmitter was in a direction 90° away from its true direction. Generally the error is less, but it may, nevertheless, be rather serious, and can be positive as well as negative.

It should not be inferred from the above that night effect is *always* present in the periods mentioned. Experience shows that only about 10 per cent of bearings taken by night at long distances have an error exceeding 10°. If conditions are such that it is necessary to take a bearing notwithstanding the appearance of night effect, it is advisable to take 10–30 bearings in the course of a few minutes and average them.

An experienced observer can nearly always discover the presence of night effect by one or more of the following symptoms:

1. Irregular displacement of the minimum.
2. When cross-bearings of three or more bearings are taken a 'cocked hat' is obtained.
3. Appearance of fading.
4. For zero sharpening, a tighter coupling is required than normally.

When it is established that night effect is present, the bearings obtained are unreliable, and it is necessary to wait till the night effect has ceased, which may take a long time. Because of this, efforts have been made to devise direction-finding systems that are not affected by night effect. A system that has been known for a long time makes use of the Adcock aerial. This aerial is constructed so that only the horizontal component of the magnetic lines of force can set up a voltage in it, and night effect no longer influences the bearing. Its dimensions for the usual frequencies are too large for installation on board ship but direction-finding stations may be equipped with it.

Coast effect

It is well known that rays of light are refracted when they pass the contact surface of air and glass. In *Figure 3.17*, A represents a source of light. The velocity of propagation in glass is less than in air and it can be proved that, by following the path ABC, the ray from A reaches the point C in the shortest time. Owing to this refraction the luminous point A is seen from C in the direction CB, at an angle that differs α degrees from the true direction.

An analogous phenomenon, which probably originates in a similar way, is found when radio waves pass a coastline. The velocity of propagation is less over land (especially in the case of a dry, sandy desert soil) than over sea, so that when a bearing is taken from the transmitting station (A in the example shown in *Figure 3.18*), a similar error α is made. When the ship–transmitting-station bearing line is at right angles to the coastline, there will be no error (as in optics), and the error increases as the angle of the line with the coastline lessens. Bearings should not, therefore, be taken from a station if this angle is small. A correction cannot be applied for coast effect; if the ship's position were, for instance, D (*Figure 3.18*) instead of C, the angle α would be different.

The error α becomes smaller if the transmitting station A is nearer the coastline. Coast effect can be neglected when taking bearings from radio beacons.

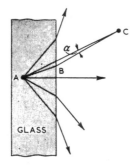

Figure 3.17 Refraction of light passing from glass to air

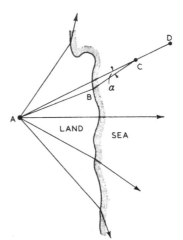

Figure 3.18 Refraction of radio waves passing over a coastline

With radio direction-finding stations, even those near the coastline, the situation is different because it is the *ship* station that is transmitting. The error is now the angle CAB (instead of ACB), which remains practically unchanged even if A is very near the coast. The sector in which coast effect does not affect accuracy, the *reliable sector*, is therefore given in the data published on radio direction-finding stations. In the worst cases, where the bearing line is nearly parallel with the coastline, 4° to 5° has been found to be the greatest error resulting from coast effect.

Other errors

In addition to the errors described, others may appear which arise partly during the propagation of the radio waves and partly in the direction-finder itself. These errors will not be discussed here, as they are usually insignificant. In addition to all these objective errors, subjective errors may occur owing to incorrect determination of the minimum signal strength and inaccurate reading of the scale and the compass. Also, any inaccuracies in the gyro-repeater or compass correction applied to the compass course pass unchanged to the true radio bearing.

PRACTICAL OPERATING INSTRUCTIONS

Often the transmitter is so weak that it cannot be heard within a sector of as much as 30°. In this case the two points on either side of the minimum, where the signals disappear and reappear, are determined. If these points are at, for instance, 17° and 25°, the average $(17° + 25°)/2 = 21°$ is the position of the exact minimum. A bearing obtained in this way is called a *swing bearing*. Swing bearings are not very accurate; one reason for this is that the zero cleaning cannot function. Therefore one should try to reduce the 'width' of the minimum as much as possible by increasing the receiver sensitivity.

The bearing with respect to the ship's head as determined by means of the direction finder is, after application of the correction, called the *relative* radio bearing. The *true* radio bearing is found by adding the ship's true heading to this relative radio bearing. When there is no gyro-compass with repeaters available on board, the procedure is as follows.

At the moment when the minimum of the figure-of-eight diagram has been adjusted, the ship's heading should be read on the compass. In order to indicate the correct moment, the operator taking the bearing presses a button and actuates an electric bell or other device near the compass as a signal to a second operator to read the heading. Before the bearing is taken a warning signal of, for instance, two dashes or dots may be given. The helmsman should now try to maintain the ship's course as exactly as possible. Then a signal consisting of one dash or dot is given at the moment when a bearing is taken. If the transmitting station sends out signals, say, for one minute, five bearings may be taken during this time, while the courses are noted and numbered by the navigating officer and the radio bearings by the observer. If the bearings are taken in quick succession, the time need be noted only once. In most cases it is not necessary to remove doubt concerning the direction by means of the cardioid diagram. If it is necessary, it need be done only once.

If one of the bearings is not reliable, e.g. because of a vague minimum, it is not used. At the end of the observations a signal of, say, three dashes is given. Then the corrections are applied to the readings. It is always necessary to check that the number of radio bearings agrees with the number of course readings. After correcting the course readings for variation and deviation, the average true course is found. This true course is added to the corresponding average radio bearing and this gives the true radio bearing. Thus the detailed procedure is:

1. Disconnect from earth all aerials except the d.f. aerial.
2. Tune the direction finder, switch it to the highest selectivity, and turn the loop or the search coil as accurately as possible to the minimum of the figure-of-eight diagram, making use, if necessary, of the zero cleaning. (We take it for granted that there is no need for determination of sense.) Read the heading at the moment of taking a bearing or press a button to warn somebody else to do so.
3. Repeat this a few times.
4. Apply the corrections and average the corrected bearings.
5. Average the courses and correct them to the true course.
6. Add the average radio bearing to the average true course and reduce the result, if necessary, by $360°$

In most cases ships are equipped with a gyro-compass. A second scale of $0°-360°$ can then be fitted round the first and the zero of this scale kept automatically in the position corresponding to true north. The true bearing can then be read off directly. It is also possible to install a repeater compass near the direction finder.

Some direction finders are provided with a second scale that can be rotated by means of a knob. If the true course of the ship is, for instance, $80°$, this scale is set so that $80°$ on it coincides with $0°$ on the fixed scale, and $0°$ on the rotating scale corresponds with true north. The true bearing can now be read on the rotating scale. In this system, however, deviations from the true course due to the ship's yawing may cause errors in the bearings.

Suppose that there is a rotatable scale (either driven from the gyro or manually set) on which the zero corresponds to true north. If the bearing reading is, say, $60°$ and the ship's course is $40°$, the transmitter bearing relative to the ship's head is $20°$. From the correction curve, therefore, *we find the correction for $20°$ and apply it to the $60°$ bearing*. This is because the amount of correction required depends on the direction of the incoming waves relative to the ship's head, and it is this correction that has been plotted in the correction curve.

As mentioned already, the windings of the loop aerial are housed in a tube. This tube must not be closed; part of its circumference is of insulating material. The insulation of the tube should never be painted, but should be kept clean and free from grease.

Conductors affecting bearings

The errors due to interfering conductors such as stays, etc., can be largely prevented by the provision of insulators. No current can then flow. As the presence of these insulators affects the corrections to be applied, they must be fitted before calibration. If they are installed afterwards, the equipment must be recalibrated. Special trouble is caused by stay eye-splices wound with rope, which are conductive in moist weather, but less so or not at all in dry weather, so that the current flowing in the stays varies. It is better, therefore, to bridge these bad contacts with a wire so that their influence on the bearings is constant. Wires of awnings may also affect radio bearings, especially if they are at the same height as the frame.

Units or sub-units of the radar installation, particularly the unit containing the modulator, must not be mounted nearer than 6 metres from the wires connecting the direction finder with its aerial, or less than 2 metres from the aerial itself if these radar units are at a greater height than the lower part of the frame aerial.

All aerials, and especially the aerials of the ship's radio station, must be disconnected from earth when taking radio bearings. In order that this may not be forgotten, there is on many ships a switch that simultaneously connects the direction finder and disconnects the ship's aerial.

Serious errors may also arise from broadcasting aerials, which are not insulated from earth, placed in the neighbourhood of the loop aerial. It is advisable to utilize a common aerial (central aerial system) for broadcast reception, this aerial being automatically switched off as soon as the direction finder is switched on.

Nowadays a Bellini-Tosi aerial is sometimes mounted in the top of a king post as far as possible from the ship's hull in order to reduce the latter's influence.

Homing

A ship may be steered so that a transmitting station continues to bear straight ahead. This is called homing, and is often employed when, for instance, approaching a lightvessel equipped with a radio beacon, or a cape or harbour with a transmitting station nearby. This method of making landfall can also be applied when the position of the ship is not exactly known, but it must of course be beyond doubt that there is no land or shallows between the transmitting station and the ship.

On the American coast a vessel collided with the Nantucket lightvessel, because homing had been continued too long and the fog signals had apparently been relied upon too much; it is, therefore, strongly recommended to steer a course that will keep clear of the lightvessel (see warning in *The Admiralty List of Radio Signals*, Volume 2). It should be borne in mind that if a bearing remains the same, a collision will take place.

Other applications of direction finders

Ships in distress can more easily be found by means of radio bearings than when the course must be set on the basis of dead-reckoning positions. The ship in distress can always be found with radio bearings, even when serious errors are made, provided that bearings are taken at regular intervals.

An important application of the direction finder is to observe the bearing of ships in the neighbourhood in foggy weather. If this bearing does not change, a collision is threatened.

If radio reception is difficult because of interference from an unwanted station (e.g., a transmitter purposely causing interference in time of war) the operator can improve reception by receiving with the aid of the direction finder (figure-of-eight or cardioid diagram) and setting the frame so that the jamming station is no longer heard. This will not succeed, of course, when both transmitters are in the same direction.

It is very important that every navigating officer who has to operate the direction finder always takes test bearings when the opportunity offers. They may be taken, for instance, when passing a cape or a lightvessel on which there is a radio beacon, or when a ship (preferably on the same course) is sighted. Radio bearings and visual bearings are then taken simultaneously. The difference between these two bearings must equal the correction. This procedure provides practice in the necessary routine. Also, repeated trials encourage confidence in the radio bearings, and at the same time the accuracy of the correction curve can be checked.

VISUAL DIRECTION FINDERS

Although the principle of all direction finders is the same, there are different methods of operation. There exists apparatus where, instead of the telephone, a cathode-ray tube or a meter indicates the strength of reception and thus the exact position of the minimum. A drawback of this is that, whereas an interfering station may be clearly *audible* in the direction of the minimum of the station from which a bearing is taken, a cathode-ray tube or a meter cannot, of course, distinguish between two transmitters and so is not usable in this case. A telephone is therefore usually available to enable a change from visual to auditive indication. In the latter case, the difference in pitch makes it possible for the ear to distinguish between the two transmitters.

Principle

Most radio-direction finders of the visual type are equipped with a cathode-ray tube, and are generally based on the Watson-Watt principle (*Figure 3.19*).

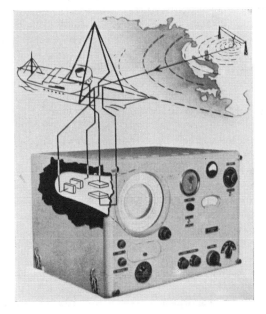

Figure 3.19 Plath visual direction finder based on the principle of Watson-Watt

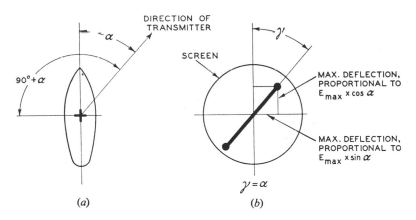

Figure 3.20 Visual direction finder: the thick line on the screen, caused by a radio transmitter, has the same direction relative to the vertical as the transmitter relative to the ship's head

It was shown earlier that the voltage induced in a loop is E_{max} cos α, where E_{max} is the voltage when the plane of the loop is in the direction of the transmitter, and α is the angle between the transmitter and an arbitrary position of the plane. Hence the voltage set up by a transmitter in the alongships frame is E_{max} cos α; thus in the thwartships frame it is E_{max} cos $(90° + \alpha) = -E_{max}$ sin α (see *Figure 3.20(a)*). The two voltages are amplified by two separate amplifiers (not shown in *Figure 3.19*) and then applied between the horizontal and vertical deflecting plates of a cathode-ray tube.

The alternating voltage set up in the alongships frame causes the light spot on the screen to move up and down. Similarly, the alternating voltage set up in the thwartships frame makes the light spot move horizontally to and fro.

The alternating voltages in the two frames are in phase, because they are due to the same change in flux. At the moment that the vertical deflection is at its maximum, the horizontal deflection will also be at its maximum. The light spot will consequently describe a straight line (see *Figure 3.20(b)* and *Figure 3.21*). It can be shown that the direction of this line corresponds exactly to the direction of the transmitter ($\gamma = \alpha$).

Figure 3.21 Bearing (without night effect) from radio beacon Lisbon

The circumference of the screen is provided with a scale graduated from $0°$ to $360°$, so that the bearing of the transmitting station from the ship's head can be read off directly. A second scale can be kept in the position corresponding to true north by means of a gyro-compass, so that the true bearing can be read off directly.

When sky waves and ground waves are received simultaneously, the voltages for the horizontal and vertical deflection are not in phase. This will result in an ellipse appearing on the screen. Since sky-wave reception is subject to rather rapid changes, the shape of the ellipse changes too (see *Figure 3.22*).

Interference by a second transmitter produces a parallelogram on the screen, as shown in *Figure 3.23*, which is a picture of the radio beacon Gibraltar (180 miles distant) jammed by the beacon Cabo Santa Maria (about 100 miles distant). The long side of the parallelogram shows the direction of Gibraltar and the short side that of Cabo Santa Maria. To

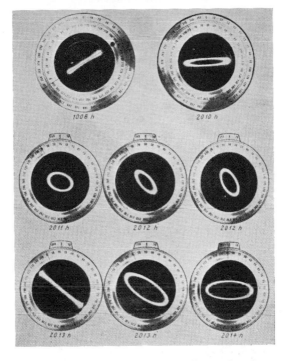

Figure 3.22 Bearings (with night effect) taken from Consol beacon Lugo during a 4-minute period

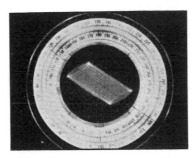

Figure 3.23 Bearing from radio beacon Gibraltar jammed by radio beacon Cabo Santa Maria

Figure 3.24 Signals from two beacons or other stations, each transmitting during the other's intervals, become visible simultaneously on the screen

facilitate the reading in such cases a rotatable glass scale provided with parallel lines is fitted in front of the screen, with the middle line extended beyond the scales.

Visual direction finders have an interesting capability. When two telegraphy stations transmit dashes in such a way that the dashes from one occur in the intervals between the dashes from the other, the signals received are not reproduced simultaneously (as in *Figure 3.23*) but alternately. Owing to the afterglow of the screen and the persistence of vision of the eye, the picture shown in *Figure 3.24* is then seen. The angle between two transmitters can be measured directly in this way, independently of the compass. When our ship is on the great circle that passes through both transmitters the two lines on the screen coincide, and so a line of leading lights can be imitated.

Balance check

In the Plath type SFP 705 LNG visual direction finder the two frame voltages are amplified separately before being supplied to the deflection plates of the cathode-ray tube. The ratio of the two voltages at the amplifier outputs must be equal to the ratio of the two frame voltages at the inputs; this necessitates equal and constant amplification. In the course of time the amplification will vary, however, and even a very small difference between them changes the direction of the electronic bearing line on the screen, resulting in a false bearing reading.

To make the amplifications equal again the inputs of the two amplifiers are switched in parallel and connected to either the alongships or the thwartships frame (see below). Because the input voltages of both amplifiers are now the same the output voltages should be the same too. If both deflection plates are supplied with equal and (we will assume) in-phase voltages, the bearing-line direction will be 45°−225°; if the amplifications are different, however, the bearing-line direction will be different (we assume that the output voltages are still in phase).

To ascertain whether the electronic bearing line has the correct position of 45°−225°, two parallel reference lines (item 11 in *Figure 3.25*) are engraved in the glass of the c.r.t. screen (not in the rotatable glass disc in front of it). If the bearing-line direction is not 45°−225°, equal amplification can be restored by means of the 'angle correction' control (item 18).

It is possible that the amount of amplification in both amplifiers is equal, but that there is a phase difference between the two output voltages. The display will then show an ellipse instead of a straight line. This can be corrected by the 'line correction' control (17). As both controls (17 and 18) are interdependent it is advisable, whenever one of them is adjusted, to verify the adjustment of the other.

Prior to each direction-finding operation the *balance check* should be carried out as described above. When using the direction finder for 'homing', a balance check should be made every 5 to 10 minutes. For balancing, the main switch (1) should be set to the position 'Bal. 1' or 'Bal. 2', whichever provides the stronger signal. The difference between these

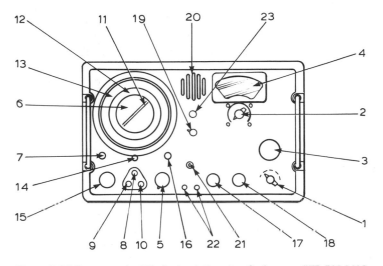

Figure 3.25 Front panel of Plath visual direction finder type SFP 705 LNG

two settings is that one connects the alongships frame to the parallel inputs of the amplifiers, the other the thwartships frame.

Switches and controls

The main switch (1) has six positions: 'Off', 'Bal. 1', 'Bal. 2', 'Dir. finding', 'Sense blue' and 'Sense red'. The mode switch (19) has three positions: 'A1' for reception of c.w. telegraphy, 'A2' for modulated telegraphy and 'A3' for radiotelephony. There are three controls for focusing and shifting the luminous spot: 'Focus' (8), 'Vertical shift' (9) and 'Horizontal shift' (10). The gyro control is 14; 21 is the loudspeaker switch and 22 is the output for the headset.

Sense determination

In visual (as in auditive) direction finders, the 180° ambiguity cannot be resolved with the figure-of-eight diagram. For sense determination the cardioid diagram is again required. For that purpose field coils are inserted in the frames of the Plath direction finder; inside these coils is mounted a rotatable search coil, connected between the sense antenna and earth; see *Figure 3.11*. In this case, however, the search coil induces a voltage in the field coils, unlike its application in auditive direction finders. The voltages in the two field coils depend on the position of the search coil relative to the field coils.

Suppose that the relative bearing of a transmitter is $0°$; no voltage will be induced in the thwartships frame. To prevent the electronic bearing line from changing direction $(0°-180°)$ the search coil should not then induce a voltage in the thwartships field coil. By means of the cursor control (15) the search coil must therefore be rotated so that it is coupled solely to the alongships field coil. To achieve this the cursor control knob is mechanically coupled to the search coil and also to the rotatable glass disc in front of the screen of the c.r.t., on which are engraved parallel lines (the cursor) to facilitate the bearing reading. Hence, by turning knob 15 so that the middle line of the parallel lines on the cursor coincides with the electronic bearing line $(0°-180°)$, the search coil is automatically coupled only with the alongships field coil.

A transmitter will generally induce a voltage in both frames, and the search coil should then induce appropriate voltages in each of the two field coils. To achieve this the cursor control must be rotated to that position where the cursor coincides with the bearing line. Because of the mechanical coupling the search coil then induces the correct voltages in both field coils. These voltages are either in phase or $180°$ out of phase with the loop voltages, making the electronic bearing line longer or shorter, respectively. The main switch should now be set to the position 'Sense red' or 'Sense blue', whichever shortens the electronic bearing line. Each switch position is marked by a coloured dot (red or blue); the corresponding colour will be found on the end of the cursor that indicates the correct bearing.

Procedure for taking a bearing

1. Disconnect transmitting antennas.
2. Set master switch (1) to position 'Bal. 1' or 'Bal. 2'; wait until luminous spot appears on screen (6).
3. Adjust brilliance control (7).
4. Adjust dial illumination (16).
5. Turn gain control (5) to extreme clockwise position.
6. Select desired frequency band (2); there are four bands: 70–140 kHz, 130–260 kHz, 250 530 kHz and 1500–3500 kHz.
7. Tune receiver to the desired station by means of tuning knob (3), referring to frequency dial (4) and then to c.r.t. display.
8. Check whether DF-signal is heard in the loudspeaker (20) or headphones.

9. Adjust audio gain by means of the potentiometer (23).
10. Change luminous spot on c.r.t. into a straight line by means of the gain control (5); make sure that the straight line does not extend beyond the c.r.t. diameter.
11. If an ellipse is shown on the c.r.t. it can be changed into a straight line by the line-correction knob (17); this can be made to coincide with the balance reference line on the c.r.t. by means of the angle-correction knob (18). For this purpose set the main switch to whichever of the 'Balance' positions makes the straight-line display longer.
12. Set master switch (1) to position 'DF'.
13. Turn cursor (15) until it coincides with c.r.t. display (i.e. with the major axis of the ellipse).
14. If sense is unknown, set master switch (1) to 'Sense blue' or 'Sense red' whichever makes the electronic bearing line shorter.
15. The relative bearing is read from the inner scale (12) at whichever end of the cursor has the same colour as the master-switch setting.
16. The true bearing may be read on the outer azimuthal scale (13).
17. Apply the correction.

Special charts

The result of the combined reception by the two frames of a visual direction finder is a signal that is always equal to the *maximum* reception (E_{max}) of the same station by one frame. This is a fundamental difference from auditive direction finders, where bearings are always taken on *minimum* signal strength. Because of this, the so-called 'maximum direction finders' can take bearings from stations at a far greater distance. Provided the receiver is suitable for the frequency of the stations, it is possible to take bearings from Loran, Consol, aeronautical and broadcasting stations at a considerable distance.

However, at long distances there is often night effect. Moreover, the error in nautical miles increases though the error in degrees remains constant. For such long-range bearings the stations involved do not always appear on the chart or, if they do, the scale of the chart is sometimes too large.

Bek/Trak Research and Development Corporation, of Elmsford, New York, have published charts with bearing lines at one-degree intervals for certain stations, e.g. Loran stations. ;

The error in the position line is, of course, proportional to the bearing error; it amounts to 1 nautical mile for a bearing error of $0.5°$ at a distance from the transmitter of 114.6 nautical miles.

All Loran-C stations use the same frequency, 100 kHz. The pulses of the stations of any chain are not received simultaneously anywhere. Therefore we see on the screen two or more bearing lines at the same time

(*Figure 3.24*). Now we can read off *the difference of two bearings* on the visual direction finder. This difference in bearing, together with the position of the two stations from which bearings were taken, suffices to determine a locus of the position of our ship. The advantage of this method is that a compass error or a yawing ship does not influence the accuracy.

The German firm Plath publishes charts (*Figure 3.26*) on which the position lines for certain station-combinations and bearing-differences are

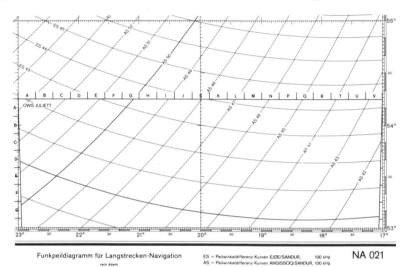

Funkpeildiagramm für Langstrecken-Navigation   ES = Peilwinkeldifferenz-Kurven EJDE/SANDUR,   100 kHz   NA 021
nach Adank   AS = Peilwinkeldifferenz-Kurven ANGISSOQ/SANDUR, 100 kHz

Figure 3.26 Example of the charts published by Plath GmbH, Hamburg; each LOP is the locus of ship's position for a particular difference in bearing from two stations

indicated. For instance, the line of position (LOP) AS 46 is a locus of the ship's position if the bearing difference is 46° for the Loran stations Angissoq (A) in Greenland and Sandur (S) in Iceland, both operating on 100 kHz.

Advantages and disadvantages of visual (Watson-Watt type) and auditive direction finders

(*a*) The indication of the bearing on visual apparatus is automatic and does not depend on the skill of the operator.

(*b*) The two amplifiers (one for each frame) in visual apparatus must be correctly adjusted before taking bearings. This is important because slight differences in amplification bring about errors in the bearings. An auditive apparatus does not have this drawback as the signals received by the two frames of a B-T aerial are combined before amplification.

(*c*) The visual apparatus is a maximum direction finder, making it possible to take bearings from distant stations.

(*d*) Atmospherics arrive from all directions. For auditive direction finders they are very strong in relation to the desired signal because the bearing signal is made as weak as possible when taking a bearing. Visual apparatuses, however, are *maximum* direction finders and the signal strength is stronger in relation to atmospherics.

(*e*) Night effect can easily be discerned with visual apparatus, because the bearing line on the screen is not straight but elliptical in shape.

(*f*) With visual apparatus it is possible to take bearings and receive Morse signals simultaneously, which is important in cases of distress.

AUTOMATIC DIRECTION FINDERS

On many ships, the traditional direction finder is often replaced by an automatic direction finder. Commercial airliners have been equipped with one or two of them for a long time.

Figure 3.27 shows the Marconi automatic ship's direction finder *Lodestar*. Simultaneous reception via a Bellini-Tosi frame aerial and an auxiliary

Figure 3.27 'Lodestar' automatic direction finder (courtesy Marconi International Marine Co.)

aerial takes place. By adding and subtracting the alternating voltage in the search coil (figure-of-eight diagram) to and from the alternating voltage in the auxiliary aerial, a cardioid diagram is obtained. Reversing the two connections of the search coil causes a phase shift of 180°. Instead of being added the voltages must now be subtracted, and vice versa; this produces a mirror image of the cardioid diagram (see *Figure 3.28*).

With the aid of an electronic switch, a phase-shift of 180° is obtained 400 times a second. Consequently the signal increases or decreases 400 times a second. After detection, this leads to an alternating current with a frequency of 200 Hz, which is fed to a small a.c. motor causing the search coil to begin to rotate. The rotation is always in the direction of the

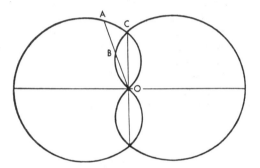

Figure 3.28 Principle of 'Lodestar': the search-coil connections are reversed 400 times a second, causing the cardioid diagram to reverse at the same frequency. Thus the strength of the received signal varies between OA and OB, for example. The resulting low-frequency current is fed to a motor that turns the search coil to position OC, where the signal strength is constant

minimum of the figure-of-eight diagram (direction OC in *Figure 3.28*). In this minimum position, the strength of signal OC is not affected by reversing the connections, and so, after detection, there is no alternating voltage. The motor no longer rotates, and the search coil remains in this position.

The figure-of-eight diagram has two minima, 180° apart. The rotation of the motor is, however, such that the search coil is driven along the shortest way to the minimum where the pointer indicates the correct bearing of the beacon, not the bearing with an error of 180°. Theoretically, it is possible for the pointer to remain on the wrong minimum, but this is an unstable position.

Naturally, automatic direction finders cannot distinguish between the signals of the required station and those of an interfering station. However, it is possible to separate these two stations by ear when taking bearings because of the difference in pitch. With the *Lodestar*, therefore, provision has been made for taking auditive bearings if necessary. Switches and controls for this purpose are arranged under the cover shown open in *Figure 3.27*.

THE ACCURACY OF RADIO BEARINGS

The accuracy of position finding by means of radio bearings depends on:

1. The presence or absence of night effect and/or coast effect.
2. The distance to the stations from which bearings are taken. (An error of 1° in the bearing means that, for every 10 nautical miles distance from the beacon, there is an error of 0.17 nautical miles in the position).
3. The angle of intersection of the bearing lines.
4. The accuracy of the correction curve.
5. The quality of the direction-finding apparatus.
6. The experience of the operators who take the bearings and read off the compass.

Bearings are only reliable if, irrespective of the amplification, the volume of the sound is zero and the sound is audible again at the slightest rotation of frame or search coil.

It may be assumed that 95 per cent of the bearings taken on board show an error of less than 3°, provided that there is no coast effect or night effect. As mentioned before, coast effect need not be feared as long as the radio beacon on which a bearing is taken is not situated too far from the coast. Night effect can occur only when the distance to the transmitting station exceeds 25–100 miles at night and 200 miles by day. Accuracy can be increased by taking several bearings and averaging them. This is specially recommended when night effect may be expected or when the ship is yawing. The presence of night effect can be inferred from a displacement of the minimum, shifting continually.

In 1950, doubts were expressed whether there was a reason for the continued existence of the direction finder, now that other position-fixing systems had come to the fore. It was argued that the technical development of the direction finder had practically come to a standstill during the previous 10 years, whereas the other systems seemed to hold so many possibilities. At present it is agreed that the direction finder is most unlikely to leave the scene in the near future.

Compared with other position-fixing systems, the direction finder, which is the simplest, still has certain advantages. The apparatus is not expensive, and though the accuracy expressed in degrees is less than that of other systems, there are many more stations available from which bearings can be taken. Consequently the average distance to the transmitting stations is less, which favourably influences the accuracy expressed in nautical miles.

Calibration and the correction curve

Except in very special cases, the determination of the correction curve and the associated adjustment of a direction finder (the calibration) is not made by the ship's personnel but by the service personnel ashore. However, the captain and navigating officers should assist. Calibration is as important for the direction finder as compensation is for the compass, and the two operations are much alike. Calibration is necessary not only when the apparatus is first installed on board, but also every time the position of any aerial or construction on deck is changed in a way that could noticeably affect the accuracy of the direction finder. It should be checked regularly.

During the calibration, the derricks, davits, etc., should be in the same condition as they will be at sea. As a rule aerials other than that of the direction finder must be isolated from earth, but there are exceptions, for instance, the Decca aerial. If an aerial is disconnected during calibration, it should also be disconnected when bearings are taken.

The presence of cranes, iron sheds, etc., in the immediate vicinity renders accurate calibration impossible.

In order to determine a correction curve, radio bearings and visual bearings must be taken simultaneously from the same transmitter. This can be done in two ways. The ship can be kept in a fixed position, while test bearings are taken from a transmitter in a lifeboat sailing round the ship at a distance of at least one mile. A better method is to swing the ship round while taking successive bearings from a fixed transmitting station. There are stations especially erected for calibrating purposes.

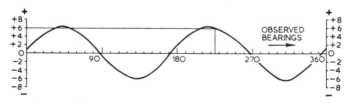

Figure 3.29 Correction curve for observed bearings

In special cases, when there has been no time for calibration in harbour, it may be done while the ship is at sea, with the aid of coast stations or other stations, but this method is rather inaccurate and cannot be recommended.

Although the correction curve should be determined for every frequency, it is usually made only for the frequency at which bearings are generally taken.

In the correction curve shown in *Figure 3.29* the correction to be applied is +6° when the bearing observed is 225°. As the correction is positive the bearing is, consequently, 225° + 6° = 231°.

RADIO BEACONS

To enable ships to take radio bearings, special transmitting stations have been established in places which are suitable from a navigational point of view, such as capes, lighthouses, etc. An idea of the number of these so-called *radio beacons* can be gained from the fact that there are eight of them along the Dutch coast.

The Admiralty List of Radio Signals, Vol. 2 (hereafter referred to as 'the list'), as well as Publication 117A and B of the Defense Mapping Agency Hydrographic Center, Washington DC and other publications, contain the necessary data about these radio beacons. An example of an entry in the list for IJmuiden Lt is shown below (by courtesy of The Hydrographer of the Navy).

0237 RC IJmuiden Lt   52° 27′ 47″ N   4° 34′ 34″ E

    294.2   A2

No.	Name	Ident	Range	Seq	Fog	Clear
0237	IJmuiden	YM	20	1, 4	Cont	Cont
0231	Hook of Holland	HH	20	2, 5	Cont	Cont
0243	Eierland	ER	20	3, 6	Cont	Nil

RC is the indication for a non-directional radio beacon for marine navigation. The following types of station also appear in the list:

RG = *radio direction-finding station* (G = Gonio).
RD = *directional radio beacon* (D = directional), i.e. a beacon that radiates only in one or more beams; often located near port entrances, fairways, etc., to be of assistance to vessels making harbour or negotiating a particular channel. The signal received varies according to the ship's position in relation to a bearing line. Bearings given in the list in respect of such stations are *towards* the station in each case. The automatic gain or volume control should be switched off when using such beacons; the ship's d.f. aerial should not be used.

On Admiralty charts, the bearing line will normally appear as a pecked line with a legend, e.g. RD046°, in red-purple (magenta). *Beam width must be borne in mind when the bearing line passes close to dangers to navigation.* In many cases, it may be prudent to borrow to one side of the beam; local regulations may in any case require ships to keep to starboard of the beam, as an anti-collision measure.

It may be found, when a directional beacon is calibrated, that the observed beam deviates from the intended bearing line along part of its length. Serious deviations are given in the list, where known; the bearing line on the Admiralty charts will normally be limited to the portion or portions in which the observed beam substantially coincides with the nominal bearing line.

RW = *rotational pattern radiobeacon.* These beacons are used only in Japan.

Calib Stn = *calibration station.*

We now continue with our example of IJmuiden. The signal from IJmuiden is, as is nearly always the case, class A2 (modulated telegraphy), because this class is technically more suitable for direction-finding purposes. The frequency is given in kHz.

In order to avoid interference, a group of two or more radio beacons often uses a common frequency (*time sharing*) that differs from the frequency assigned to neighbouring beacons or groups of beacons. Beacons of the same group do not transmit simultaneously but in immediate succession. IJmuiden cooperates in this way with Hook of Holland and Eierland.

Each station in a group is listed at the correct place in the geographical sequence, but details of the group are given only in the table appearing under the station allocated the sequence number 1 (Seq. 1). At each other station in the group the user is referred to the list number of the station giving the table.

The sequence number indicates the order of transmission in the cycle. The system now in general use is based on a unit of 1 minute of time and a cycle of 6 minutes; in this system, a common frequency is shared by a group of up to six beacons. In all cases the group cycle is taken as commencing at the hour, *being repeated every 6 minutes thereafter.* Each beacon in the group transmits at a fixed time within the group cycle, as indicated by the sequence numbers 1 to 6. The table opposite shows the relationship between sequence numbers and times of transmission; the figures below the sequence numbers indicate the time of commencement of transmission, in minutes past the hour.

When the group frequency is shared by fewer than six beacons, various arrangements are possible. For example, IJmuiden has the sequence numbers 1 and 4, so it transmits during the first and fourth minutes of each group cycle; Hook of Holland transmits during the second and fifth minutes, etc.

1	2	3	4	5	6
00	01	02	03	04	05
06	07	08	09	10	11
12	13	14	15	16	17
18	19	20	21	22	23
24	25	26	27	28	29
30	31	32	33	34	35
36	37	38	39	40	41
42	43	44	45	46	47
48	49	50	51	52	53
54	55	56	57	58	59

Note that in Japan and South Korea time-sharing is not based on a 6 minute period.

Identification signal

Nearly all radio beacons transmit a Morse identification signal ('Ident' in the list). The identification signals of such stations are included in the station details and are also indexed at the back of the volume.

Certain short-range *marker radio beacons* are not covered by the Alphabetical List of Identification Signals, but in view of their short range there should be no difficulty in identifying them in practice.

Radio beacons operating on radio-telephony transmit an identification signal consisting of the name of the radio beacon in speech; such signals are self-evident and are omitted from the Alphabetical List.

Some directional radio beacons do not transmit an identification signal in addition to the directional signals. Being of short range they should present no difficulty as regards identification.

QTG service

When there are no suitable radio beacons in the neighbourhood the radio officer can ask certain specified coast stations to function temporarily as a radio beacon by transmitting for one minute. This can be requested by the code signal QTG; the service is, therefore, called a *QTG service*. The coast stations of this QTG service are also included in the list. Normally a small amount is charged for these special transmissions.

There are also radio beacons that transmit on request in order to enable ships to calibrate their direction finder.

Taking bearings

To determine the position we have to proceed as follows:
1. Look up on the chart suitable radio beacons, paying attention to a favourable intersection of the bearing lines.
2. Note from one of the lists particulars such as the frequency, identification signals, and time of transmission.
3. Take the bearings by proceeding as listed on page 110.

Synchronization of radio and sound signals for distance finding

The distance to a lightvessel or other site can be determined by making use of synchronized radio and sound signals. This method is especially important during fog.

The speed of propagation of sound is 330 m/s or 330/1852 = 0.18 n. mile/s, so the distance to the sound source in nautical miles is 0.18 times the travelling time of sound (in seconds). A travelling time of 1.1 second indicates a distance of 1.1 × 0.18 = 0.2 nautical miles approximately. The travelling time of radio signals may be neglected.

The *beginning* or the *end* of a *time signal* is synchronized with the *beginning* or the *end* of a *blast of an air fog signal*. Details of this 'timing point' can be found in the service details of each station employing this method. A stop watch should be started on reception of the radio 'timing point' and stopped at the instant that the air fog signal 'timing point' is heard. The elapsed time is the time taken for the sound to travel through the air from the air fog signal site to the ship. The distance off, in nautical miles, is obtained by multiplying the elapsed time, in seconds, by 0.18.

A second method is to transmit a series of radio 'measuring signals'. The first of these commences simultaneously with the beginning of the air fog signal timing point. Distance off is determined by noting the number of intervals between measuring signals received before the air fog signal timing point is heard. The distance represented by each interval is given for each station. When, for instance, the interval between the commencement of the successive signals is about 1.1 second, the unit of distance will be 0.2 nautical miles. Thus the use of a stop watch is not required.

The air fog signal in the US consists of a blast of 1 second, a silence of 1 second and a blast of 5 seconds. The radio signal is radiated at the commencement of the long blast of 5 seconds. The blast of 1 second acts only as a warning before the timing point.

Radio direction-finding stations

A ship, whether equipped with a direction finder or not, can have a bearing supplied by *radio direction-finding stations*. These stations, established at fixed places, take bearings of ships on radiotelegraphic request. The bearings are then communicated to the ship.

Most direction-finding stations are better equipped technically than ships and are as free as possible from interfering conductors in the vicinity. For this reason, and also because the compass error on board is not known very exactly, the bearings furnished are in general more accurate than those taken on board.

Radio direction-finding stations are either equipped with transmitting and receiving apparatus or connected by land-lines to a transmitting and receiving station; the latter can be a normal coast station. In the last case, requests for a bearing to be taken by one or more direction-finding stations should be addressed to the coast station. This station advises the direction-finding stations concerned by telephone-line. The ship then has to transmit signals for about one minute, after which the bearings are transmitted by the coast station to the ship.

Radio direction-finding stations supply *true* bearings of the ship as observed from the direction-finding station. The general regulations applying to the use of these stations are laid down in the *Manual for use by the Maritime Mobile Service*, published by the International Telecommunications Union, Geneva.

Aero radio beacons

A number of radio beacons for air navigation are included in the list of radio beacons. They are located so that they can also be used by ships, and are marked in the list by the letters Aero RC. The advantage of these beacons and of the *radio ranges* (which are directional aero radio beacons) is that they mostly transmit continuously. It is very important to bear in mind that the inclusion of an aero radio beacon in the list does not imply that it has been found reliable for marine use. It is not possible to predict the extent to which coast effect may render bearings unreliable.

4

The Consol system

The Consol system is a hyperbolic position-fixing system with a baseline of only about 2½ nautical miles. It can be assumed that, at a distance from the mid-point of the baseline exceeding about twelve times the length of the baseline, each hyperbola coincides with its asymptote (the asymptote is a straight line through the mid-point of the baseline); see *Figure 1.41*. Hence, except for distances shorter than about 30 nautical miles from the mid-point of the baseline, we may consider the hyperbola as a straight line and (on the surface of the Earth) as a great circle passing through the mid-point of the baseline. The system should not be used at distances of less than about 30 nautical miles from the transmitter.

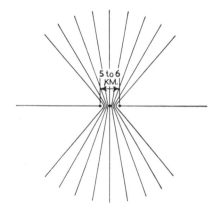

Figure 4.1 The short baseline of the Consol system ensures that (except near the transmitter) the hyperbolas are virtually straight lines; the three points on the baseline are the aerial positions

Just as a radio direction-finder provides the bearing of a radio beacon, so the Consol system provides the bearing of the mid-point of the baseline. See *Figure 4.1*. Hence a Consol transmitter, though part of a hyperbolic system, is called a Consol *beacon*.

A Consol beacon has three aerials placed in á straight line (the baseline); there is only one transmitter, which is connected to the three aerials. By gradually changing the phase of the alternating currents in the two outer

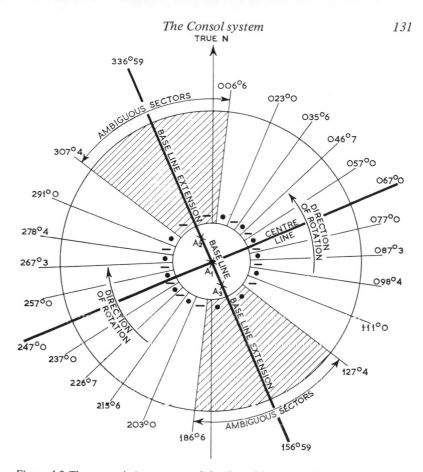

Figure 4.2 The transmission pattern of the Consol beacon at Stavanger; A_1, A_2, A_3 are the three aerials. At the start of each transmission, dots are heard in 12 sectors, dashes in 12 sectors. The pattern rotates so that each sector has moved into the position of the next in 30 seconds; the rotations on either side of the baseline are opposite in direction

aerials with respect to that in the centre aerial, directional transmissions are obtained.

The coverage of most Consol beacons is divided into 24 sectors. *Figure 4.2* shows the transmission pattern of the beacon at Stavanger in Norway. In 12 of the sectors dots are heard at the commencement of the cycle of operation, and in the other 12 sectors dashes are heard, as indicated in the diagram by a dot or a dash. The sectors are not all of the same size: those near the baseline extension are largest, those near the centreline are smallest. Dash sectors alternate with dot sectors, except that the two sectors on either side of each baseline extension are the same. In each sector, 60 signals (dots and dashes) are heard during one cycle.

When the observer is on the limit between a dot and a dash sector he hears the dots as well as the dashes; because the dots are emitted in the intervals of the dashes, and vice versa, the result is a continuous tone, called the *equisignal*.

A characteristic of the radiation pattern is that it is not stationary: the transmission pattern on one side of the baseline in *Figure 4.2* rotates to the left, that on the other side rotates to the right. This takes place in such a way that at the end of the keying cycle an equisignal has replaced the next equisignal. Thereafter the transmission stops for a short time and, after some preliminary non-directional signals from the centre aerial alone, the procedure is repeated.

The number of sectors, the duration of the transmission and the direction of the baseline are not the same for the various Consol beacons.

Reading Consol signals

In *Figure 4.3*, OA, OB and OC represent lines on which the equisignal is heard at the commencement of the dot-dash cycle. The ship is located

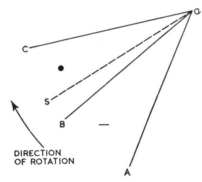

Figure 4.3 *If an observer in sector BOC hears, say, 15 dots and 45 dashes he is on line OS, because angle BOS is 15/60 of the sector angle BOC*

somewhere on the line OS. Sector AOB is a dash sector and BOC a dot sector. The equisignal line, which at the outset had the position OA, has reached the position OB at the end of the 30-second dot-dash period (Stavanger), while in the meantime the line which was in position OB has moved to OC. In this time 60 signals (dots and dashes) are transmitted. The angular velocity at which the equisignal lines rotate may be assumed to be constant.

While, therefore, the equisignal line OB is rotating towards OS, dots are heard on board; then, during the rotation from OS towards OC, dashes are heard. If the equisignal is heard after counting 15 dots, for instance, the radius OB has rotated over an angle BOS, equal to 15/60 times the sector angle BOC. The bearing lines for the various sectors are printed on special charts, showing the number of dots and dashes that are heard on each line before the equisignal; by counting the dots or dashes before the signal, the observer knows which line in this sector is the ship's position line.

Identifying the sector

It is not possible to obtain data in this way about which sector the ship is in. Even the smallest sectors, however, subtend an angle of about $10°$ so identification affords no great difficulties. In most cases, the dead-reckoning position will eliminate possible doubt, but the Consol transmitters also lend themselves very well to the taking of bearings with the direction finder. For this purpose, a prolonged dash is transmitted by the centre aerial before the start of dot-dash transmissions. *Figure 4.4* shows the

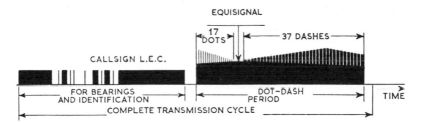

Figure 4.4 Stavanger signal strength. During the dot-dash period, 17 dots are heard followed by the equisignal and 37 dashes; the middle of the equisignal falls at the twentieth dot, so the observer's position line is the Consol line for 20 dots within the appropriate sector

signal from Stavanger, where for identification of the Consol beacon the callsign LEC is inserted. This callsign is transmitted slowly to enable anyone to identify the beacon.

The equisignal

Unfortunately the transition from dots to dashes or from dashes to dots is very vague. *Figure 4.4* shows the signal strength during a typical complete cycle of a Consol beacon. It will be seen that the intervals between the dots appear to be filled up with dashes, which at first are much weaker and are not heard. These dashes, however, gradually become stronger and the dots weaker. After the 17th dot there is no longer a contrast between the dots and dashes, and a continuous tone, the beginning of the equisignal, is heard. A short time later the dashes are heard more strongly than the dots, which are transmitted in the intervals. In total 37 dashes are counted. Now it may be assumed that the missing signals consist half of dots and half of dashes. Since 60 signals in all have been transmitted and in this case 17 + 37 = 54 have been counted, there are 60 − 54 = 6 missing, 3 of which are dots and 3 dashes. The true count is therefore 17 + 3 = 20 dots and 37 + 3 = 40 dashes. As the dots were heard first, we are in a dot sector.

Assuming that we know which dot sector we are in, we only have to look on the chart for the bearing line in this sector with the indication 20 dots. This is shown as 20·. On the bearing line for 20 dashes the indication would be 20−.

At present the Stavanger dot-dash period (*Figure 4.4*) has a duration of half a minute and the whole cycle lasts one minute.

As is evident from the foregoing, not only the characters before the equisignal but also the characters following it should be counted. When 25 dashes are counted before the signal and 31 dots after it, four signals have been missed and the true count is 27 dashes and 33 dots; we are thus in one of the dash sectors.

It may also happen that the equisignal is heard immediately, followed by 56 dashes, after which the transmission ceases. The number of missing signals is then four, two of which are dots and two dashes, and the true count is two dots and 58 dashes. In this case the two missing dots were received before the middle of the equisignal and we are, therefore, in a dot sector.

Consol and Consolan beacons

Near the baseline extensions the system cannot be used, for two reasons:
1. The accuracy is very poor in those regions.
2. Adjacent to each baseline extension there are two similar sectors, e.g. two dot sectors.

There are at present (1977) 13 Consol and Consolan beacons. Operationally there is no difference between Consol and Consolan. Particulars can be found in *The Admiralty List of Radio Signals*, volumes 2 and 5. For each beacon the list gives: position of the middle of the baseline, frequency, callsign, period of dot-dash cycle, keying cycle and hours of operation. The beacons are: Stavanger and Andoya in Norway, Bjornoya on Spitsbergen, Jan Mayen Island, Plonéis in France, San Francisco in the United States, Lugo in Spain and six beacons in Russia.

By day the range is 500 to 1000 nautical miles; at night a still longer range can be expected.

Reception from stations

A great advantage of Consol is that it needs only a normal receiving apparatus suitable for the reception of continuous waves at frequencies of about 300 kHz. In most cases such an apparatus is already on board. Lifeboats can also easily be equipped with a receiver suitable for Consol.

For reliable counts, however, a good selectivity is required. To increase the selectivity, some receivers are equipped with a crystal filter which can be successfully used for Consol. Only the very narrow frequency band of the Consol beacon transmitters can pass through this filter. Atmospheric interference and interference from neighbouring frequencies are reduced, as a result of which it is easier to count the dots and dashes.

Tuning

Most receivers are equipped with automatic volume control (a.v.c.); this control should of course be switched off when Consol signals are received, as otherwise the contrast between dots and dashes would be insufficient for them to be distinguished.

The receiver should consequently be adjusted as follows:

1. Switch A1/A2/A3: this should be set to A1 (as we must be able to receive continuous waves).
2. Switch for automatic volume control: off.
3. Switch for frequency-band and tuning control: to frequency of the particular Consol transmitter.
4. Selectivity switch: in the position for the highest selectivity.

When the receiving apparatus is not sufficiently selective, much interference is experienced from normal radio beacons operating on neighbouring frequencies. A direction finder is suitable for reception in such cases. If it has a rotatable loop, this should be turned so that it is at most 45° distant from the position for maximum sound.

For tuning, some receivers are equipped with a so-called 'magic eye', which can be valuable in determining the centre of the equisignal, so increasing the accuracy of the count. There are also receivers in which a meter indicates the signal strength of incoming signals.

Accuracy

The accuracy of all Consol stations is greatest on the centreline, mainly because the sectors are smallest there. The larger the bearing with respect to the centreline, the less the accuracy, and, as pointed out, the system must not be used near the baseline extensions. This is a shortcoming to which all hyperbolic systems are more or less subject.

By day over sea the 95 per cent error in the count on the centreline is two dots or two dashes, corresponding to one-third of a degree. At an angle of 60° from the centreline, the 95 per cent error expressed in signals remains the same, but expressed in degrees it is twice the previous value, and on the baseline extensions it can even be twenty times as much.

By night another factor comes into play, because reflection by the ionosphere will appear, and at distances between about 300 and 500 nautical miles both the direct and the reflected radiation is received. This gives rise to further errors. The accuracy in degrees is greater at distances from the transmitter that exceed 500 nautical miles, because then only the reflected radiation is received. In practice, however, the navigator on board is more concerned with the accuracy of the position, and this depends not only on the foregoing but also on the angle of intersection of the two position lines.

By making several successive counts and taking the average of the results, the number of variable errors decreases (the systematic errors remain the same). This procedure is especially advisable at night.

The accuracy of Consol position fixes, compared with Loran and Decca fixes, is not high. In fact, Consol beacons are normally seldom used on merchant vessels. In a Notice to Mariners, a warning was given against the use of the Consol system. This notice reads in part: 'Although the system is useful for ocean navigation, Consol gives bearings which are insufficiently accurate for landfall or coastal navigation, and mariners are warned that these bearings cannot be relied on with safety when closing danger. For example, at Spurn Head it is possible for bearings from the Stavanger Consol station to give a position line as much as 12 miles in error at night. By day the probable error will be smaller than by night but the same limitations of the use of the system apply.'

The accuracy of the Consol system, expressed in degrees or signals, is consequently:

1. Better near the centreline than near the baseline extension.
2. Better by day than by night.
3. Better over sea than over land.
4. Poorer by night at distances between about 300 and 500 nautical miles than at shorter and longer ranges.

The accuracy in miles will of course become poorer as the distance from the beacon increases.

Under normal conditions, Consol bearings are much more accurate than bearings obtained by means of a direction finder. Thus with Consol by day and at an angle of 60° to the centreline, a 95 per cent error of two-thirds of a degree has to be allowed for, whereas the 95 per cent error of a bearing taken with the direction finder is about 3°.

On the other hand there are many more radio beacons than Consol stations; the distance to the beacons is, therefore, usually much shorter, and this increases the accuracy. So where there are radio beacons at short range they are used for preference. The choice depends also on other circumstances, for instance the direction of the position line obtained. The direction of the centreline of Plonéis is such that ships approaching the Channel are in an area of favourable accuracy.

When it is not possible to obtain an astronomical or other position, the Consol position may serve to check the dead-reckoning position.

Consol charts and tables

When we have completed the count and know in which sector we are located, we can immediately look up the Consol line on the chart. This is the simplest and most accurate method. On most charts the Consol lines are printed not for every individual signal but for every 5 or 10 or sometimes more (see Chart I). This depends also on the portion of the chart that is needed. It will, therefore, generally be necessary to interpolate, which can be done with sufficient accuracy by eye.

On the charts, the lines of the various beacons are clearly differentiated by colour. At present many British, French and German Consol charts are available; Admiralty Consol charts are normal sea-charts overprinted with Consol lines.

The number of signals for the sectors near the baseline extensions is not 60 but less. In order to prevent the system from being used here, the Consol lines may be omitted in the most unfavourable portions of these sectors; this is done, for instance, on the Admiralty Consol charts. On other charts such sectors are marked as 'ambiguous sectors'.

Consol charts are not corrected and must not, therefore, be used for navigational purposes to the exclusion of other charts. Latitude and longitude of the position on the Consol chart have to be measured and transferred to the normal sea-chart.

If no Consol chart is available of the area in which the ship is sailing, the Consol position line can be plotted with the aid of tables. These tables are included, for example, in *The Admiralty List of Radio Signals*, volume 5. For each sector the true bearing (great-circle bearing from the beacon) for each whole number of signals is shown in them.

Plotting position with the aid of Consol charts

An example of a Consol position-fix with the aid of Consol charts can now be given. We are sailing by night in the Atlantic. Position by dead reckoning is 49° 33' N, 7° 3' W (see S on Chart I).

First we tune in to Stavanger (319 kHz). After hearing the call sign LEC and the prolonged dash we count 44 dashes, the equisignal and 8 dots. The number of counted signals is 52, so that half of 8 dashes, that is 4 dashes, are missing. The true count is, therefore, 44 + 4 = 48 dashes. As it is night we repeat this count a few times. We now look up on the chart the Consol line marked 48— that is nearest the dead-reckoning position.

Then we tune in to Plonéis (257 kHz) and count 13 dots, the equisignal and 41 dashes. The true count here is, therefore, 16 dots. This count is also repeated a few times as a check, after which we plot this line; in doing so we have to interpolate (see Chart I).

As our Consol position we find 49° 35' N, 6° 58' W. This position is transferred to the normal chart.

5

The Decca system

Among position-fixing systems for navigation, the Decca system holds an important place. This is because the position of the receiving station can be determined with a high degree of accuracy.

On the transmitting side ashore there are four stations (called a chain) that transmit continuously, while on board a special receiving apparatus is required. This receiver is connected to three indicators (decometers). In order to distinguish between them they are called the red, green and purple decometer. The words *red, green* and *purple* appear on the respective dials, or, on recent models, each dial is marked with a small appropriately coloured rectangle.

Each indicator gives a reading that refers to a definite line on the map. The position of the ship lies on this Decca line. Three lines of position can be obtained in this way, but two of them are sufficient to determine the position.

PRINCIPLE

In *Figure 5.1*, A and B represent transmitters both of which generate continuous waves. The frequency of A is kept accurately constant. Station B is fitted with a receiver tuned to the transmitter of A, and connected to the transmitter of B in such a way that the current in the transmitting aerial of B has a constant phase difference with that of A. For convenience we will assume the phase difference to be zero, i.e. the aerial currents of A and B are in phase. A is called the *master* and B the *slave*. AB is the *baseline*.

In a receiving aerial at any point, say P_1, of the centreline, two alternating voltages are set up by the radiation from A and B. As the distance $P_1 A$ is equal to $P_1 B$ and the currents in the transmitting aerials are in phase, it is obvious that the two voltages at P_1 are also in phase.

At a point a short distance to the right of P_1 the distance to A has increased and that to B decreased. Consequently the voltages will no longer be in phase. The further the point moves to the right of P_1, the greater the phase difference will become, and at P_2 this difference is $180°$.

Still further to the right at P_3, the phase difference has increased to $360°$; in other words, the zero-phase ($0°$) condition is regained. The distance AP_3 is then one wavelength longer than the distance BP_3. A difference in distance of one wavelength corresponds to a phase difference of $360°$

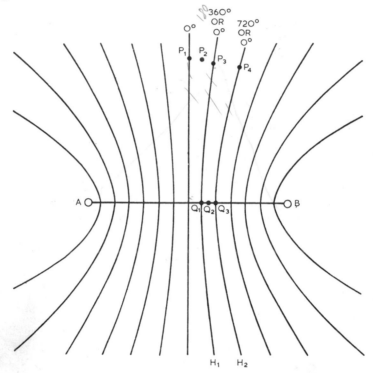

Figure 5.1 The alternating currents in transmitting aerials A and B are in phase. At P_1 the alternating fields are in phase. If P_3 is such that $AP_3 - BP_3 =$ one wavelength, the alternating fields at P_3 are $360°$ out of phase, i.e. also in phase. This applies to all points on the hyperbola through P_3. From P_1 to P_3 the phase difference thus changes from $0°$ to $360°$; similarly, if $AP_4 - BP_4 = 2$ wavelengths the fields are again in phase on the hyperbola through P_4

A *hyperbola* is a line on which all points have a constant difference in distance from two fixed points, the *foci* (see page 35). If we draw the hyperbola with the foci A and B through P_3, it follows that any of its points is further from A than from B by one wavelength. At all points on this hyperbola the two alternating voltages in the receiving aerial are therefore in phase. If P_4 is two wavelengths further from A than from B, the phase difference between the two waves or radiated fields is again zero, and this holds good for all points on the hyperbola through P_4.

The area between two adjacent hyperbolas with a phase-difference of $0°$ is called a *lane*. The whole area of *Figure 5.1* can be divided into lanes.

Finding the position of a ship

If a ship proceeds from some point on the hyperbola through H_1 to any point on the hyperbola through H_2, the phase difference will change steadily from $0°$ to $360°$, i.e. $0°$ again.

By means of the Decca apparatus it is possible to measure the phase difference between the two fields. When this difference is known, the hyperbola on which the ship is to be found can be deduced from it, at least as long as the ship remains in the same lane. The intersection with another locus, for instance another hyperbola found with the aid of another master and slave, will determine the position of the ship.

Up to now we have called the locus a hyperbola. This would be so if everything took place on a flat plane. In reality the lines, which show small deviations from a hyperbola, are spheroidal hyperbolas, or spherical hyperbolas if we conceive of the Earth as a globe. They are computed and printed on nautical charts. In future we shall call them *Decca lines*.

From the above it will be clear that the Decca system is a hyperbolic position-fixing system.

Frequency conversions

Hitherto it has been assumed that, although A and B radiate at the same frequency, the two signals are received separately and their phase difference is measured by the Decca apparatus. This is in fact not possible; if two high-frequency alternating fields have the same frequency, the two alternating voltages set up by them in a receiving aerial will combine into one voltage, and it will be impossible to separate them in order to measure their phase difference. Therefore the method shown in *Figure 5.2* is not possible; the two 340 kHz signals of master and slave, which are induced simultaneously in the receiving aerial on board and which are different in phase, cannot be separated at point Q. The achievement of the American inventor of this system, W.J. O'Brien, is that he has nevertheless made measurement of the phase difference possible.

Before the principle is explained, the following remarks should be noted. In telecommunications, an alternating current, an alternating voltage or an electromagnetic field is usually designated simply by its frequency, this being the most obvious characteristic. Now a frequency of an alternating voltage or current can easily be changed into its double,

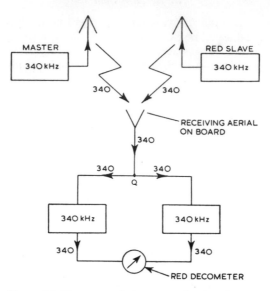

Figure 5.2 If master and slave transmitted at the same frequency (e.g. 340 kHz) the two alternating voltages in the receiving aerial would combine into one. They could not be separated at Q, so the system is impracticable

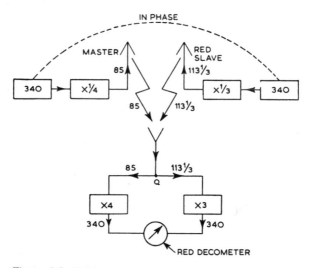

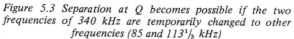

Figure 5.3 Separation at Q becomes possible if the two frequencies of 340 kHz are temporarily changed to other frequencies (85 and 113¹⁄₃ kHz)

triple, etc. It is also possible to change a frequency into one that is exactly one-half, one-third, etc., of the original. This is the basis of the method utilized in the Decca system.

The method of *Figure 5.2* is, as has been said, unusable. Therefore the two frequencies of 340 kHz that have to be transferred from the two transmitting stations to the receiver on board are first converted into the sub-multiple frequencies, 85 and 113¹/₃ kHz (*Figure 5.3*), and are transmitted in this form. By using selective circuits tuned to 85 and 113¹/₃ kHz, it is possible to separate the two frequencies at Q in ᵗhe receiver, so

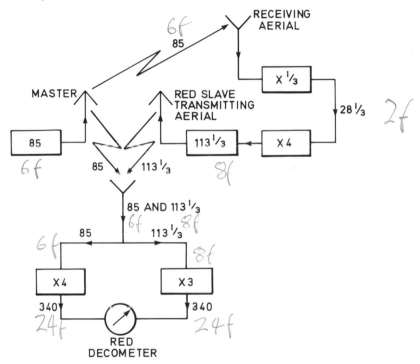

Figure 5.4 In a practical system, to ensure that the transmission from the slave has the correct phase relationship with that from the master, the slave receives its basic 85 kHz input from the master

that they can be separately converted back to 340 kHz. The phase difference between these two 340 kHz alternating voltages can then be measured, to obtain the result that was impossible in the *Figure 5.2* system.

Figure 5.4 shows how the combination of master and red slave operates in practice. The master frequency is 85 kHz instead of 340. This lower frequency is received by the slave, which is equipped with a receiving aerial and a receiver. The latter is connected to the transmitter via frequency dividers and multipliers, so that the phase of the alternating current in the transmitting aerial of the slave is not arbitrary but has a certain relation (*phase-locked*) to the phase of the transmitting aerial current of the master.

The purpose of the frequency conversions is simply to meet the requirement that the two frequencies should differ on their way from the transmitters to the receiver, so that they can be separated in the receiver. Similar frequency conversions, but to different values, take place in the green and purple slaves; the frequencies are 255 kHz for green and 425 kHz for purple.

Synchronization of the slave aerial transmitting current to that of the master can be achieved without the use of radio waves if atomic oscillators are used in both transmitters. As mentioned on page 10, these oscillators have such a high frequency stability that they make continuous synchronization by radio waves superfluous. The Decca Navigator Co. employs rubidium atomic oscillators for chains where the propagation paths between master and slave stations are very long or very bad, e.g. in the Bombay and the Calcutta chains and in some Norwegian chains.

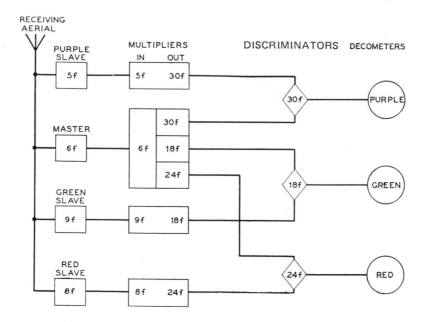

Figure 5.5 Block diagram of receiver. Frequency 6f originating in the master is multiplied by five, three and four to give the comparison frequencies 30f, 18f, and 24f. Frequencies 5f, 9f and 8f originating in the purple, green and red slaves are multiplied to the same comparison frequencies. The phase difference between each pair of comparison frequencies is measured in the discriminators and displayed on the decometers

In the slave of *Figure 5.4* the frequency of 85 kHz is divided by three, multiplied by four and then radiated by the transmitting aerial. The red slave transmits therefore at a frequency of $85 \times \frac{1}{3} \times 4 = 113\frac{1}{3}$ kHz. The frequency conversions in the receiver are indicated in *Figure 5.5*. All the frequencies are multiples or *harmonics* of a certain frequency f $(14\frac{1}{6}$ kHz), called the *fundamental value*. The master frequency $(6f)$ is therefore $6 \times 14\frac{1}{6} = 85$ kHz, and is changed in the multiplier into $30f$

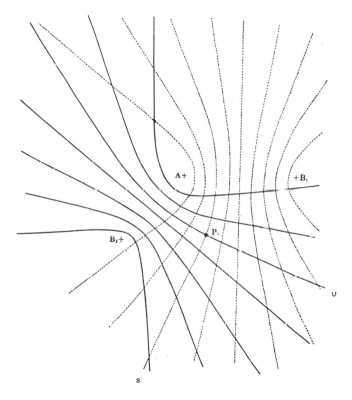

Figure 5.6 Master A controls three slaves, only two of which (B_1 and B_2) are shown. Thus three sets of Decca lines (two shown) are obtained, identified by the colours red, green and purple. If a ship is on Decca lines u and s, her position is P_1

(425 kHz), $18f$ (255 kHz) and $24f$ (340 kHz). One of the discriminators measures the difference in phase between the two 'comparison frequencies' $24f$ of master and red slave; this difference in phase is indicated by the red decometer. The other discriminators measure the difference in phase between the comparison frequencies $30f$ (purple) and $18f$ (green).

The Decca chains

To determine a second locus of the ship's position, a third transmitting station B_2 supplies a second set of Decca lines in conjunction with *the same master* station A, so the master performs a double function (see *Figure 5.6*). A second decometer responds to the transmissions from the two stations A and B_2. When the ship is, for instance, on the Decca line u in *Figure 5.6* from the pair of stations A and B_2, and at the same time on the line s from stations A and B_1, the ship's position is the point of inter-section P_1 of the two lines.

To obtain angles of intersection that are as favourable as possible, a fourth station in conjunction with the same master gives a third set of Decca lines. The four stations that form a chain are located so that the three slaves are approximately at the points of an equilateral triangle, with the master at its centre. The distance from the master to each slave is 60—120 nautical miles.

It is customary to use only two of the three decometers for position-fixing operation, i.e. the two that will give the greatest accuracy in the area concerned. Data sheets issued by Decca (see later) indicate the decometers to be used in various areas.

Figure 5.7 shows the coverages of the Decca chains existing or under construction in 1977, and the table opposite lists the chains.

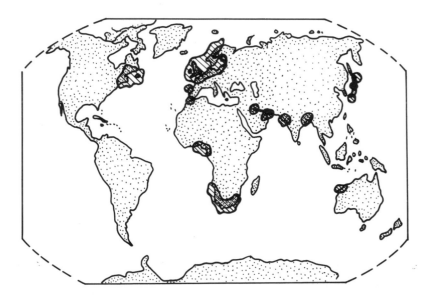

Figure 5.7 Coverage of Decca chains in 1977

Chain	Desig-nation	Chain	Desig-nation
South Baltic (Sweden)	0A/MP	Tohoku (Japan)	6C/MP
Vestlandet (Bergen,		North Scottish	6C/MP
Norway)	0E/MP	Gulf of Finland	6E/MP
South West British	1B/MP	Danish	7B/MP
South Persian Gulf	1C/MP	Bombay (India West)	7B/MP
Northumbrian	2A/MP	Nova Scotia	7C/V
*South Eastern (Nigeria)	2B/MP	Kyushu (Japan)	7C/MP
East Newfoundland	2C/MP	Irish	7D/MP
Holland	2E/MP	Finnmark (North Cape,	
*Salaya (India)	2F/MP	Norway)	7E/MP
*Mid Western (Nigeria)	3A/MP	*Rivers (Nigeria)	7F/MP
North British	3B/MP	Eastern Province (S.	
Lofoten (Norway)	3E/MP	Africa)	8A/MP
German	3F/MP	French	8B/MP
Port Hedland (W.		Calcutta (India East)	8B/MP
Australia)	4A/MP	South Bothnian	8C/MP
Namaqua (South Africa)	4A/MP	Dampier (W. Australia)	8E/MP
North Baltic	4B/MP	Hebridean (W. Scotland)	8E/MP
North West Spanish	4C/MP	*Lagos (Nigeria)	8F/MP
Trondelag (Trondheim,		Frisian Islands	9B/MP
Norway)	4E/MP	South West Africa	9C/MP
English	5B/MP	Anticosti (Canada)	9C/MP
North Persian Gulf	5C/MP	Hokkaido (Japan)	9C/MP
North Bothnian	5F/V	Helgeland (Polar Circle,	
Cape (South Africa)	6A/MP	Norway)	9E/MP
*South Spanish	6A/MP	Skagerrak	10B/MP
Cabot Straits (Canada)	6B/MP	Natal (South Africa)	10C/MP
Bangladesh	6C/MP		

*under construction

Lane width

For the sake of clarity the lanes in *Figure 5.1* have been drawn far too
wide in relation to the baseline length: the total number of lanes may be
200 or more.

The width of a lane depends on the comparison frequency. In *Figure
5.1*, Q_1 is the point of intersection of the Decca line H_1 and the baseline.
The phase difference here is $0°$. Let us now take on the baseline a point
Q_2 that is a quarter wavelength nearer to B. (Note that the wavelength

referred to here is that which corresponds to the *comparison* frequency, not the frequency at which the station transmits. The point at issue is the phase difference between the two comparison frequencies, as measured eventually in the apparatus. We may, therefore, for this description, assume that master and slave both operate at the comparison frequency.)

The waves from A arrive later at Q_2 than at Q_1 by the amount of time they take to travel a quarter wavelength. As the field travels a distance of one wavelength during the time corresponding to one cycle, the time required for travelling through a quarter wavelength corresponds to a quarter of a cycle, or $90°$. The field from A at Q_2, therefore, lags $90°$ behind the one at Q_1. It will be clear that the field from B at Q_2 *leads* the one at Q_1 by $90°$. Since the fields from A and B were in phase at Q_1, they are $90° + 90° = 180°$ out of phase at Q_2.

If Q_3 is another quarter wavelength nearer to B than Q_2, the two fields at Q_3 will have a phase difference that is $180°$ greater than at Q_2 and $360°$ or $0°$ greater than at Q_1.

All points of the Decca line H_2 have, therefore, a phase difference of $360°$ from all points of H_1. The area limited by H_1 and H_2 is consequently a lane and its width measured on the baseline is half a *comparison wavelength*. If this wavelength is, for instance, half a nautical mile and the distance between the stations is 50 n. miles, there are $50/\frac{1}{4} = 200$ lanes.

Normal and 'reference' operation

Figure 5.8 shows a simplified block diagram of the Decca Navigator Mark 21 receiver for both normal and 'reference' operation. This diagram and its description can be omitted by those readers who are not interested in electronics.

The purple slave frequency $5f$ is amplified, and thereafter controls a phase-locked or 'flywheel' oscillator (see pages 11–12), which thus automatically acquires the frequency and phase of the purple slave. The oscillator output is free from noise and is less subject to atmospherics and man-made noise.

The oscillator frequency $5f$ is then multiplied by six to $30f$ and supplied to the purple discriminator. (The red and the green slave frequencies $8f$ and $9f$ are converted in the same way to $24f$ and $18f$ respectively, but this is not shown in the diagram.)

The master frequency $6f$ controls the frequency and phase of another flywheel oscillator. The oscillator frequency is multiplied by five to $30f$. and then supplied to the purple discriminator (as well as to the green and red discriminators) where the difference in phase between the master and the purple slave frequencies is measured. The purple decometer indicates the result in centilanes and (by the counting action of the geared dials) in lanes and zones.

The 6*f* flywheel oscillator is also connected, via a 90° phase-shifter, to a *cosine discriminator*. The discriminators hitherto referred to are called *sine discriminators*; as described in Chapter 1, they produce a positive or negative error voltage that depends on the phase difference between the received frequency and the oscillator frequency. The error voltage is continuously supplied to the oscillator, and alters the frequency and phase of the generated signal in such a way as to bring the phase difference down

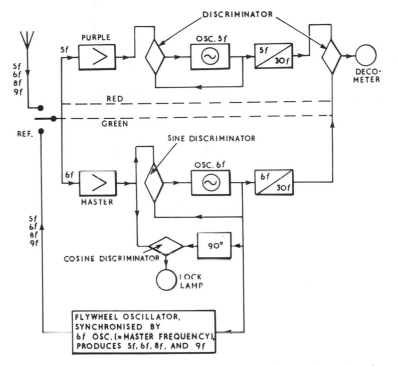

Figure 5.8 *Mark 21 normal and 'reference' operation. In normal operation the 6f and 5f flywheel oscillators acquire the phase and frequency of the master and purple slave respectively; when the 6f oscillator has locked on to the correct phase, maximum voltage is produced in the cosine discriminator and the 'Lock' lamp lights. In reference operation the signals received from master and slave are replaced by reference signals*

to zero; the error voltage itself thereby becomes zero. In the case of the cosine discriminator, however, the error voltage is brought to a *maximum* as the phase difference reaches zero; in the Mark 21 circuit, it is this voltage that causes the 'Lock' lamp to illuminate.

A third application of the 6*f* flywheel oscillator is to synchronize a fifth flywheel oscillator (not indicated in the diagram). When the function switch is set to 'Ref.', all four frequencies (5*f*, 6*f*, 8*f* and 9*f*) are obtained from this last oscillator by multiplication and division, and are supplied to the rest of the circuit as a substitute for the signals received via the aerial. The purple, red and green discriminators should then detect zero phase difference and all three centilane pointers should go to zero. If one or more does not, the operator should adjust the appropriate 'Zero' control(s) (see page 169) so that all the pointers read zero.

Decometers

There are three types of Decca receiver. The oldest is the Mark V, followed later by the Mark 12; at the end of the 1960s the Mark 21 was introduced. Only the decometers of the Mark 21 will be discussed in detail.

The Mark V and 12 decometers each have two pointers, coupled together, to indicate the lane and centilane, and a rotating disc coupled to the lane pointer to indicate the zone, through a window in the dial.

The Mark 21 decometers, however, each have only one pointer (see *Figure 5.23*). When the ship has crossed one lane, the phase difference between the received signals has increased by 360° or one cycle and the pointer has made one revolution; its scale is marked in hundredths of a cycle or centicycles (one cycle = 360°, one centicycle = 3.6°), to indicate centilanes. The width of successive lanes is not the same, except along the baseline; the same holds true for centilanes, but this is neglected when plotting the position on the chart. One should be aware, however, that the position line for a lane-reading of, say, 21.50 is not exactly halfway between the lanes 21.00 and 22.00.

In the Decca Operating Instructions the word 'lane-fraction pointer' is used, but in accordance with Omega terminology we shall keep to 'centilane pointer' or 'cel pointer'.

The centilane pointer is coupled via gearing to a disc. For the red decometer, to which we shall confine ourselves, the disc has a scale with the numbers 0 to 23 at its edge. The disc turns behind the dial, with one (sometimes two) of the numbers visible through a window. When the ship has crossed one lane the centilane pointer has made one revolution (100 cel) and the disc has turned from one number to another on its scale.

In its turn the lane disc is coupled to a second disc or ring, visible through the same window, from which the zone can be read. When the lane disc has made one revolution the ship has crossed one zone.

There are 10 zone letters, A—J, starting from the master station; see *Figure 5.9*. Some baselines are longer than 10 zones and the A—J lettering is therefore repeated. Zones with the same letter identification are too far from each other to cause confusion.

The number of lanes in a zone differs for each decometer; there are 24 for the red indicator, 18 for the green and 30 for the purple. The 24 lanes of a red zone are numbered from 0 to 23, the 18 green from 30 to 47 and

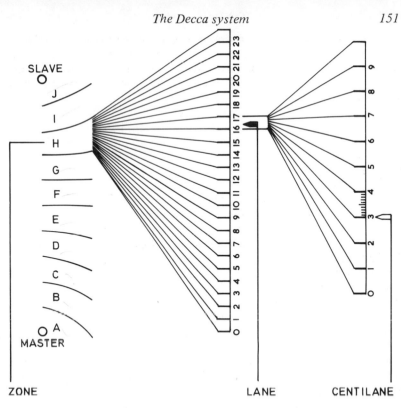

Figure 5.9 Red pattern divided into zones, lanes and centilanes; the reading in this case is red H 16.30

the 30 purple from 50 to 79 (see table). This is done in order to avoid confusion: a position fix can be given (e.g. in a distress message) without having to indicate the colours. The lanes are numbered within a zone from master station to slave.

Decometer	Zone letter	Lane number	Centilane number
Red	A–J	0–23	0–99
Green	A–J	30–47	0–99
Purple	A–J	50–79	0–99

Lane-slip

Because the lane disc is coupled mechanically to the electrically-driven centilane pointer, the lane reading, once set up, will remain accurate while the ship is under way.

When the ship leaves the region inside which the Decca system may be used, the apparatus should be switched off; when the ship later re-enters Decca coverage at another point and the apparatus is switched on, the centilane pointer will set itself to the right position but the lane reading will not be correct.

This also occurs if the receiver on a ship navigating within Decca coverage has been switched off for some time. Therefore the apparatus should be kept in continuous operation while in a Decca area. Even when the ship is in the vicinity of a baseline (e.g the Port of London) and moves some distance with its apparatus switched off, an altered lane indication should be taken into account.

Another danger is that there may be a temporary breakdown of one or more transmitters, or the transmissions may be incorrect, resulting in incorrect lane indication. In such cases a notice to mariners to this effect is transmitted as soon as possible by radio telegraphy or radio telephony. The failure or malfunctioning of a transmitter is, however, very improbable, as experience has shown. In such a case, a stand-by transmitter automatically takes its place, and a third transmitter is switched into stand-by duty within 1½ seconds.

Even when the transmitters are functioning normally an erroneous lane can be indicated (so-called *lane-slip*). It is possible, for instance, for a signal received as a ground wave to have a $180°$ phase difference from the same signal received as a sky wave; if the waves are equally strong, one will neutralize the other and there will be no reception. So although the ship continues her voyage the centilane pointer stops turning. Suppose that this occurs at a reading of 11 lanes and 12 centilanes (11.12), and that when the signal is received again (after some time) the actual position is 11 lanes and 88 centilanes. The centilane pointer will set itself to the correct new position by going the shortest way from 12 to 88, i.e. *back via the zero*, producing a reading of 10.88 instead of 11.88.

Simultaneous reception of ground and sky waves can cause lane-slip in another way. A sky wave, having followed a longer path than a ground wave, differs in phase from the latter. If conditions are such that the sky wave increases in strength, its influence on the phase-difference measurement will become greater and the centilane pointer will give a false reading. (This may occur not only on a moving ship but also when anchored.) Ground-wave reception is stable, but sky-wave reception is sometimes irregular and may suddenly weaken or cease altogether. In the latter event, the phase-difference of the signals received will change suddenly and the centilane pointer will return at once *along the shortest way* to the correct position. If, solely as a result of gradually increasing sky-wave reception, the pointer had moved more than half a revolution away from its correct position, it will therefore complete that extra revolution if the sky wave suddenly ceases. The *centilane* reading will then be correct but the *lane* reading will be one lane too high or too low.

Lane-slip occurs very rarely. If the lane reading is erroneous for any reason, and the navigator unwittingly continues to rely on it, there may be serious consequences, however. The reading can be checked, of course, if the ship's position can be determined by other means; but this is not always possible, and a system dependent upon such checking would in any case lose much of its value. An improvement of the system was therefore needed so that the lane reading could be checked. This checking is called *lane identification* (LI).

LANE IDENTIFICATION

On page 147 it was shown that, at the baseline, the width of a lane is equal to half of the comparison wavelength. If, at regular intervals, the system operates for a short period (say 0.5 second) at a lower frequency, so that a much longer comparison wavelength is obtained in the receiver, the lanes will widen considerably. In this way one obtains, besides the normal fine pattern, a coarse pattern (see *Figure 5.10*). The comparison frequency is reduced so that a 'wide lane' is equal to a zone in normal transmission.

Inside a lane the Decca receiver indicates accurately which centilane the ship is in. This holds good for the 'wide lanes' too: the receiver calculates which part of a 'wide lane' (i.e. which part of a zone) the ship is in; thus it can make certain which normal lane the ship is in. As a rule there is no doubt about the zone, since the width of a zone subtends an angle of about 12° at the mid-point of the baseline.

Since a red zone is equal to 24 red lanes, the lane must become 24 times as wide at red LI. The comparison frequency of 340 kHz must therefore become 24 times as low, i.e. $340/24 = 14^1/_6$ kHz.

Multipulse lane-identification

The frequency of $14^1/_6$ kHz can be obtained in two different ways. The first method, used in the Mark V equipment, although still in use, has almost disappeared and is therefore not discussed here. Moreover, the latest type of receiver, the Mark 21, is not suitable for this type of LI.

In the new system for obtaining $14^1/_6$ kHz, called Multipulse, three sequences of LI signals are transmitted each minute. A sequence consists of a short Multipulse transmission from each of the four stations in turn at 2½ second intervals (*Figure 5.13*). For the master transmission, all three slaves interrupt their transmissions for a short time and only the master station transmits at four frequencies simultaneously, i.e. at all frequencies at which the four stations are normally transmitting. The receiver on board amplifies the four frequencies of the master station, and

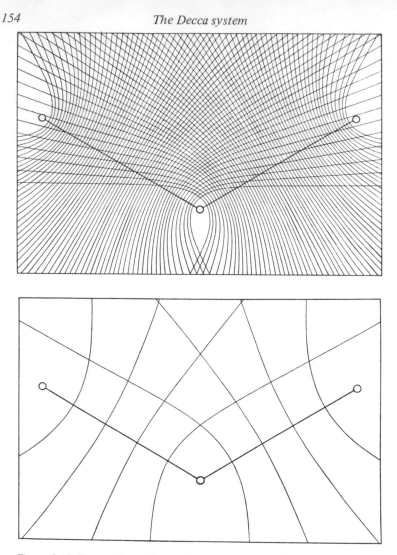

Figure 5.10 Fine pattern (above) gives great accuracy but also greater possibility of lane error, so coarse pattern (below) is used to eliminate ambiguity

they are then combined. From these four frequencies, one alternating voltage of $f = 14^1/_6$ kHz is obtained. See *Figure 5.11*.

The frequency $6f$, generated in the flywheel oscillator of the Mark 21 receiver, has been synchronized during *normal* transmissions by the master frequency $6f$ and thereafter divided by six. A measurement is now made of the difference in phase between this frequency f and the frequency f obtained from the four frequencies of the LI master transmission.

After analogue-to-digital conversion, the result of the phase-difference measurement is shown in three digits on a decimal numerical display. The reading should be 00.0. If it is not 00.0 or near to this, one should adjust it by means of the 'LI zero' knob. The phase of the master has now been transferred to the oscillator; because of its great stability the flywheel oscillator retains this phase for some time.

After the LI transmission from the master, the four stations resume sending their normal frequencies. During the next LI signal in the sequence, it is the turn of the red slave to transmit on all four frequencies, and from this the receiver again obtains a frequency f. The numerical display now

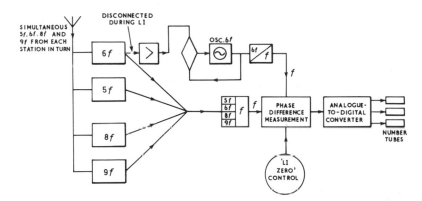

Figure 5.11 Lane indication. Oscillator frequency 6f is synchronized, during normal transmission, by master frequency 6f. The four frequencies received during an LI transmission are converted to frequency f, which is compared in phase with the oscillator frequency (after division by six). LI transmission from the master should produce a reading of 00.0 on the digital display; LI transmission from a slave produces (owing to the phase difference) a reading of lanes and tenths of a lane

indicates the difference in phase between the LI master oscillations and those of the red slave, in lanes and tenths of a lane; this gives the red position line of the ship.

The green and then the purple LI transmissions take place after further intervals of 2½ seconds.

The advantage of the Multipulse method for obtaining the frequency f is that a variation in phase and/or magnitude of one or more of the four received frequencies (e.g. as a consequence of sky-wave reception) has no great influence on the phase of the frequency f. This is very important, as it increases the accuracy and reliability of lane identification under night conditions at the longer ranges.

Figure 5.12(a) shows the four frequencies arriving in an LI transmission from one station, viz. $5f$, $6f$, $8f$ and $9f$. Since the four frequencies are multiples of f ($14^{1}/_{6}$ kHz) their resultant can be represented by a periodic curve, i.e. a curve repeating itself continuously. *Figure 5.12(b)* shows the 'added' four frequencies; the oscillation with the greatest amplitude can be separated (*Figure 5.12(c)*), and from this a sine-wave of $14^1/_6$ kHz can be obtained (*Figure 5.12(d)*).

The system so far described provides:

(*a*) a fine pattern, with readouts on decometers, and with $24f$ (340 kHz) as the comparison frequency (for simplicity's sake only the red slave is being considered).

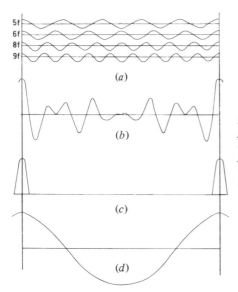

(*a*)

Figure 5.12 Multipulse: (a) the four frequencies transmitted simultaneously by one station; (b) the four frequencies added; (c) pulses obtained from highest peaks of (b); (d) sine wave finally obtained, frequency f

(*b*) a coarse pattern, with f ($14^1/_6$ kHz) as the comparison frequency, and lanes that are 24 times wider than in (*a*), there is a digital readout (on number tubes) in lanes and tenths of a lane. Simultaneous transmission of four frequencies by each station in turn makes the phase of the resulting frequency less dependent on sky-wave effects.

Time schedule for LI signals

Decca chains are classified as types V1, V2 and MP, according to their method of LI transmission. Of approximately fifty chains at present in existence, only two are type V1 or V2. Since it is intended to change even these to type MP, we shall describe this method only.

Figure 5.13 shows the time schedule of Multipulse (MP) signals. At intervals of 2½ seconds the master and the red, green and purple slaves in turn transmit at four frequencies simultaneously. The digital readout flashes on and off to identify the master signal. With chains that still transmit LI signals of the V2 type at the same time, there is a fifth signal, which can be disregarded with Mark 21 equipment. The sequence is repeated every 20 seconds (see *Figure 5.13*).

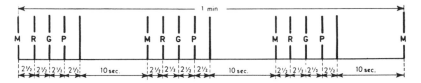

Figure 5.13 Three Multipulse (MP) LI sequences; the fifth pulse can normally be neglected

Where a chain has only two slaves (e.g. the Holland chain), there is no LI signal for the missing colour.

Advantages of Multipulse

Decca charts are based on the assumption that only the ground wave is received. If the received signal is composed of ground and sky waves, the measured phase-difference will no longer accurately indicate the ship's position line.

The undesirable influence of sky waves particularly disturbs the LI of type V1 and V2 chains. In general the effect of sky waves becomes unacceptable at distances of more than 240 nautical miles from the centre of a Decca chain; for this reason this distance has been officially accepted in some areas as the nominal limit of reliability of Decca.

As mentioned before, the advantage of Multipulse LI is that a phase shift of one or more of the four transmitted frequencies, as a consequence of sky-wave reception, has little influence on the phase of the fundamental frequency derived from the combined signal. Thus, with Multipulse chains, lane-identification is possible at greater distances, or in other words the sky wave can be of greater strength before LI becomes unreliable. The LI reading can, under such circumstances, be more accurate than the decometer readings, despite the fact that the latter imply a greater reading accuracy due to their scale divisions in hundredths of a lane.

By day, LI ranges of 400 nautical miles or more are obtained; at night the range may be expected to be 250 nautical miles under good reception conditions, but sometimes no more than 200 nautical miles, depending on conditions. For example, in regions near the magnetic equator the night range tends to be reduced to the east and west of a chain.

It is now possible to obtain a position line at long distances solely by LI readings. By means of the chain-selector controls one can switch to another chain, from which within one minute another position line can be determined in the same way as before. In some regions these position lines give a good intersection.

Most Decca charts have the disadvantage that the Decca lines of only one chain are indicated, so when proceeding in this manner the position line must be transferred from one chart to the other.

FREQUENCIES

The frequencies at which the English chain (the oldest chain) operates are: master 85 kHz, red slave $113\frac{1}{3}$ khz, green slave $127\frac{1}{2}$ khz and purple slave $70\frac{5}{6}$ khz.

Because of the high selectivity of Decca receivers and the distance between the chains, it is possible for other Decca chains to have frequencies

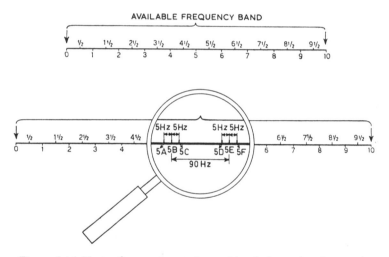

Figure 5.14 Master frequencies: stations with whole-number frequencies have B in their designation, those with frequencies halfway between have E; on either side of the B and E frequencies there is room for further ones designated A, C, D and F

differing by only a few hertz. (The ratios between the frequencies of master and slaves must of course remain the same for every chain.) In many parts of the world, the frequency band in which Decca transmitters operate has been allocated for Decca use and one need not fear interference from other transmitters.

The Decca frequency band contains eleven equally-spaced principal frequencies, indicated by the whole numbers in *Figure 5.14*. All the chains operating with these as their master-frequencies have the letter B in their designation. Then there are so-called 'half' frequencies, giving a further ten master-frequencies at ½, 1½, 2½, etc. The chains using these are indicated by the letter E.

Although there is room in this way for 10 + 11 = 21 chains, frequencies for more chains were nevertheless required. They have been found by selecting new master-frequencies on either side of each original frequency, with a very small frequency difference. These frequencies are indicated by A, C, D and F (*Figure 5.14*).

Thus a total of 63 master frequencies is available and, as each Decca chain consists of four stations, there is a total of 252 frequencies. There are currently some fifty chains in navigational use or under construction (see page 147).

ACCURACY

To get the best results it is important that the causes of errors are understood by the user of the Decca system. The various types of error are discussed below.

Fixed errors

In some areas, where the signals pass over ground of low conductivity, the hyperbolic patterns have small distortions. These distortions have been determined for certain coastal regions, and in the Decca data sheets the corrections to be applied for these fixed errors are shown on charts. There are separate correction charts for each chain and for each colour. *Figure 5.15* shows the corrections, in centilanes, for the fixed errors in the red pattern of the SW. British chain.

Only rarely does the error exceed half a lane, although there are some corrections of 0.8 lane and more. A circle around a correction indicates that it is negative and therefore must be subtracted from the Decca reading; corrections without a circle are positive and must be added.

Fixed errors may also be present in areas where they have not yet been measured and are therefore not indicated. When considering what corrections to apply in those areas one should proceed carefully. Sometimes it is possible to interpolate or extrapolate; in other cases the number of indicated corrections is too sparse to justify this, and one should not then rely too much on the accuracy of Decca.

One should also bear in mind the possibility of small inaccuracies in the charts. Some hydrographic authorities, but not all, issue charts on which the Decca lines have been corrected for fixed errors. Opinions are divided as to the wisdom of presenting lattice charts in this form.

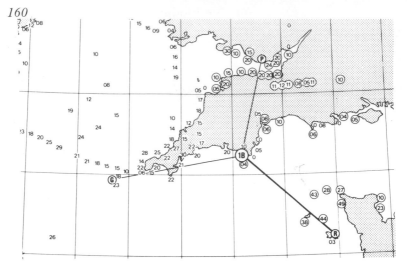

Figure 5.15 Chart of fixed-error corrections, in centilanes, for red pattern of SW. British chain; figures circled should be subtracted, figures not circled should be added

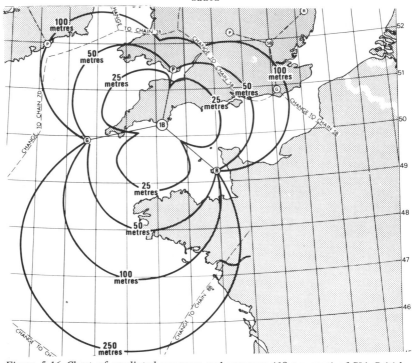

Figure 5.16 Chart of predicted coverage and accuracy (68 per cent) of SW. British chain in full daylight: the thick lines enclose areas in which 'fix-repeatability' errors will not exceed the distance shown on 68 per cent of occasions in full daylight

Variable errors

Variable errors arise mainly from the simultaneous reception of ground and sky waves. Sky-wave reception increases with distance from the chain. The phase of the resultant received waves then varies randomly with respect to the ground waves upon which the charts are based. Ground waves and sky waves can even cancel each other to some extent; a very weak resultant signal is then received, or no signal at all. The strength of the sky-wave signal is in general reduced by day; at night and at long distances, however, its effects may be considerable.

The accuracy of measuring differences in phase, indicated in centilanes, is nominally the same everywhere in the coverage. However, position-fixing accuracy increases as the baseline is approached, since the centilane width decreases and the angle of intersection of the Decca lines improves.

Data about variable errors appears in the Decca data sheets and in *The Admiralty List of Radio Signals*, volume 5. The chart in *Figure 5.16* has been taken from one of the data sheets. It shows the predicted 68 per cent accuracy in full daylight of the SW. British chain. This means that the

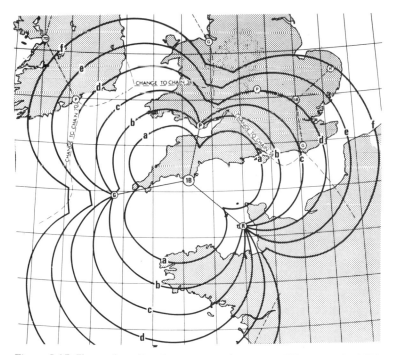

Figure 5.17 Chart of predicted coverage and accuracy (68 per cent) of SW. British chain in conditions other than full daylight; to be used in combination with Figure 5.18

accuracy, in metres, of a position on one of the curves of *Figure 5.16* is in only *one out of three cases, on average*, more than the number of metres indicated for that curve. Near Ouessant, for instance, the accuracy is 50 metres.

If there is no 'full daylight' in terms of sky-wave activity, another chart should be used; see *Figure 5.17*. The accuracy in miles on the contours a, b, c, etc. can be found in a table (*Figure 5.18(a)*) for half light, dawn/dusk,

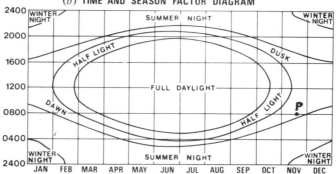

(*a*) **RANDOM FIXING ERRORS AT SEA LEVEL IN NAUTICAL MILES 68% PROBABILITY LEVEL**

DECCA PERIOD See Time and Season Factor Diagram below	CONTOUR					
	a	b	c	d	e	f
HALF LIGHT	‹0·10	‹0·10	‹0·10	0·13	0·25	0·50
DAWN/DUSK	‹0·10	‹0·10	0·13	0·25	0·50	1·00
SUMMER NIGHT	‹0·10	0·13	0·25	0·50	1·00	2·00
WINTER NIGHT	0·10	0·18	0·37	0·75	1·50	3·00

Figure 5.18 Decca period corrections. For example, on 15 February at 1200 local time near Portsmouth (between curves a and b in Figure 5.17) it is 'half light'; according to the table, the accuracy is then 0.10 n. mile or better

summer night and winter night, as defined in the 'time and season factor' diagram (*Figure 5.18(b)*). It was mentioned previously that sky-wave reception, and consequently accuracy, depends on light/dark conditions. Thus in *Figure 5.18(b)* it will be seen that, at 0800 local time on 20 November (point P) the dawn condition applies. According to *Figure 5.18(a)* the predicted accuracy at Le Havre (inside contour d) for variable errors is 0.25 nautical miles, or better, at dawn.

Charts and tables such as *Figures 5.15* to *5.18* and *5.21* are inserted in the data sheets for every chain as the data become available. It is strongly recommended to use them, particularly in coastal areas and areas that are dangerous from a navigational point of view.

Weather effects

Caution must be exercised, and frequent lane checks carried out, under weather conditions of rain and snow, since the performance of LI and decometers can be affected by this.

Regular position plotting

Accuracy at long distance and at night can be considerably improved by plotting the position on the chart at short intervals (15 minutes or one hour, depending on circumstances). After a while, assuming that the course has not changed, one may draw on the chart a straight line that goes as near as possible to all the Decca positions; in this way variable errors are to some extent eliminated. This procedure is recommended as a precaution against unexpectedly large errors (e.g. lane slip); by plotting, such errors can practically always be detected.

At a long distance from a chain the Decca lines have a small intersection angle, which reduces the accuracy; see *Figure 5.19*. The shape of the

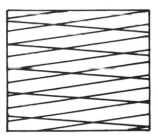

Figure 5.19 At a long distance from the chain the intersection angle is small. Accuracy is still good, however, in the direction perpendicular to the longer diagonal of the parallelogram

parallelograms between the Decca lines shows, however, that the accuracy in a direction perpendicular to the long diagonal is still quite good. Use can be made of this by approaching, for instance, a lightvessel from a predetermined direction.

Multichain charts

Formerly the UK Hydrographic Department published *interchain fixing charts* on which the Decca lines of two chains overlapped one another.

This made it possible to use the Decca lines of one chain in combination with those of the other chain, to give a better intersection angle than would have been possible using the Decca lines of a single chain.

In the past five to ten years, however, several new Decca chains have been added to the West European complex, namely the Frisian, Irish, Holland, Northumbrian and Hebridean chains. Consequently the former weak areas of Decca coverage, where the interchange fixing technique was advocated and employed, have disappeared. The UK Hydrographic Department therefore withdrew the interchain fixing charts.

The *multichain charts* of the L(D-MC) series are not a replacement for the interchain fixing charts. As a general rule, they are small-scale charts covering large areas encompassing the coverage of three or more Decca chains. The policy for such charts is to draw the lattices for the individual chains only as far as the recommended changeover line, without any overlap with the lattice for the adjacent chain. Such small-scale charts are primarily used for route planning and look-ahead, rather than for immediate navigational plots of the ship (which are performed, of course, with the aid of the largest-scale chart available for the area, invariably a single-chain latticed chart).

DECCA CHARTS

Decca charts are ordinary charts overprinted with Decca lines. They are produced and issued by at least twelve countries. With few exceptions (see above) the lines of only one chain are shown on a given chart; see Chart II.

After having applied the correction for fixed errors, if available, the position obtained by the decometer readings is plotted on the Decca chart. The longitude and latitude can then be read off and the position transferred to the ordinary chart. This is, however, normally not necessary. British, Dutch and most other charts are kept up to date through 'Notices to Mariners', and are sold corrected to the date of acquisition by the user; this date should be marked on the chart by the agent's stamp. The charts should thereafter be kept up to date in accordance with 'Notices to Mariners'. This saves the need to transfer plots to a separate updated navigational chart.

For each chain a list of Decca charts with particulars is published in the Decca data sheets and in chart catalogues. Mariners are cautioned that the larger-scale charts must invariably be consulted.

The numbering of British Admiralty charts is preceded by the letter L (lattice) followed by the letter D (Decca) and the number of the chain in brackets. So, for example, L(D7)3761 stands for Admiralty chart no. 3761 on which the Decca lines of chain 7 (Danish chain) are shown.

The complete designation of the chain is printed in the bottom left-hand border. First the frequency indication appears, e.g. 9B, then MP or V and finally the name of the chain; e.g. 'Chain 9B/MP (German)'.

When plotting on the chart, interpolation between Decca lines can be effected by eye or with the aid of a special transparent ruler with increasing scales (*Figure 5.20*). It is preferable to indicate the Decca position always in the same way, for example by the symbol △.

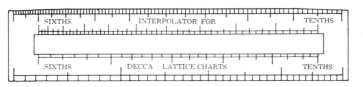

Figure 5.20 Transparent interpolator rule for Decca charts

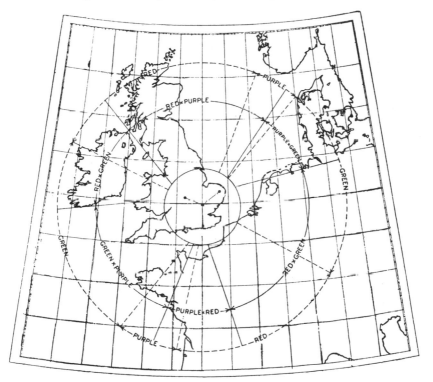

Figure 5.21 Chart showing the two colours giving greatest accuracy in different sectors of the coverage of the English chain. Within the inner circle, choice of colours will be obvious from inspection

If the position in Decca coordinates is transmitted to another ship or coast station it is necessary to include the number of the chain and also latitude and longitude.

Depending on the lane width in the particular area, a Decca line may be printed for every half-lane, lane, two lanes, or more. Near the Decca stations the curvature of the lines is apparent, but at a great distance the lines appear to be straight. As mentioned before, the Decca system has been approved up to 240 nautical miles distance from the master. By way of warning all lines at this distance are interrupted.

Under normal conditions only two of the three decometers (colours) are used. The two with the greatest accuracy for a given area can be found from special diagrams in the Decca data sheets. For each chain there is such a diagram, for example *Figure 5.21*. Inside the small circle, near the baseline, the two colours to be used can be chosen by the navigator; in doing so he should take into account the fact that accuracy increases as the distance between the lines decreases and as the angle of intersection becomes greater. For the sectors delineated by solid 'spokes', the two colours to be used are as shown. In the sectors delineated by broken-line spokes, the single colour is indicated that gives the greatest accuracy for one position line.

MARK 21 DECCA RECEIVER

A basic difference between the Mark 21 receiver and its two predecessors is the very high selectivity obtained by the use of flywheel oscillators in the various frequency channels. These oscillators generate noise-free replicas of the incoming signals. For this reason the apparatus continues to function correctly under conditions of atmospherics or electrical interference. Another difference with former types is the better Multipulse LI digital readout and the ease of IF. Also there is only one pointer on each decometer; its predecessors had two on each decometer and two on the LI indicator.

Whereas there are 21 crystals in the Mark 12 there is only one in the Mark 21. A 'synthesizer' produces from this crystal all frequencies required for the oscillators in the four channels.

Owing to advances in microelectronics, reliability has increased and the power required has been reduced to only 15 voltamperes a.c. Servicing is simplified by the use of plug-in modules; see *Figure 5.22*.

The standard aerial is an insulated wire of 6 to 10 metres; for special purposes there is a shorter aerial, insulated with plastic, that is connected to the receiver via a built-in preamplifier.

Front panel

The Mark 21 (*Figure 5.23*) has a display with a control panel below it, accessible by opening a cover. There are the following displays, switches and other controls.

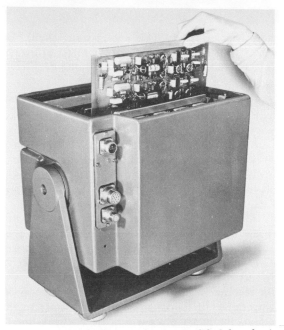

Figure 5.22 Mark 21 servicing is simplified by plug-in modules (courtesy Decca Navigator Co. Ltd)

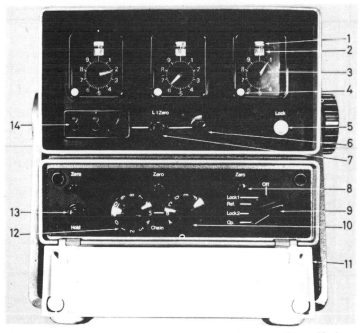

Figure 5.23 Front view of Mark 21 receiver (courtesy Decca Navigator Co. Ltd)

Chain switches. The left-hand chain switch (12) should be set to the *number* and the right-hand switch (10) to the *suffix* of the required chain. In *Figure 5.23* the receiver is tuned to chain 5B, the English chain.

Function switch. This switch (9) has five settings:

Off. This disconnects the receiver from the power supply.

Lock 1. A discriminator forces the flywheel oscillator to assume gradually the frequency and phase of the incoming signal ('locking'). The receiver should first be switched to 'Lock 1' if it has not been in use for a long time. When the receiver has only been switched off for a short time, or when switching-on under high-noise conditions, locking may take less time at the 'Op.' or 'Lock 2' settings. 'Lock 1' is only necessary if neither of these settings gives satisfactory locking.

Ref. Basically the Decca Navigator receiver measures phase differences. Thus, between the aerial and the decometers where the phase measuring takes place, the equipment must introduce no change in the phase difference. Owing to temperature variations the receiver does not completely meet this requirement. Phase-difference alteration that arises in the receiver is called 'drift'. As in any other electrical apparatus the temperature of the receiver will rise after switching on; after about one hour the temperature will become steady, and thereafter the drift is only very slight. When the receiver is switched to 'Ref.' (reference) the aerial is disconnected and the signals are supplied instead by a special oscillator. The phase of these signals is such that, if there is no drift, the difference in phase indicated by the meters will be zero, i.e. the centilane pointers will indicate zero on their scales. If they do not, each decometer can be adjusted to zero by its own 'Zero' control (8). Referencing may be required during the first hour after switching on, and after changing chains.

Lock 2. This setting ensures that the receiver re-locks as quickly as possible, after referencing, to the incoming signals.

Op. This is the setting for normal operation.

'Lock' indicator lamp. After locking, this lamp (5) illuminates steadily except for a flicker at each LI reading. The lamp gives an indication of signal/noise conditions, and illuminates brightest when these conditions are good. Flickering will increase with the level of noise. The lamp may be dimmed by turning its amber cover.

'Hold' pushbutton. When this button (13) is pressed the three centilane pointers stop rotating and move to a nominal reading of 12.5 centilanes. As a consequence of the mechanical coupling, the lane and zone indications of the decometers also become fixed. This control is useful for IF, to prevent the loss of the stored lane and zone information as a consequence of the rapid rotation of the centilane pointers. The 'Hold' button must also be pressed when switching from 'Op.' to 'Ref.'. When initially switching on, it can be pressed until the lock lamp illuminates steadily.

Decometers. The red (left), green (middle) and purple (right) decometers indicate the centilane (3), lane (2) and zone (1), the latter two through a window. A reset knob (4) on each meter allows the lane and zone dials to be initially set by hand, after which the dials are driven automatically by gearing from the centilane pointer.

Below the window there is a dot to the left and a dash to the right; there is also a dash or a dot just below each lane number inside the window. When two zone letters appear together in the window, the correct letter is the one positioned over the same symbol as that displayed under the lane number. In *Figure 5.24*, for instance, below the lane number 3 *inside the*

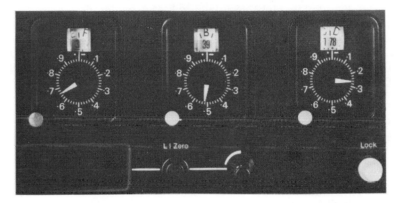

Figure 5.24 View of Mark 21 decometers showing dots and dashes to decide correct zone reading

window of the red (left) decometer there is a dash, indicating that the correct zone letter is F because F appears above the dash *below the window*. The reading is therefore Red F 3.66. In the purple (right) decometer there is a dot below the lane number 78; the zone letter above the dot on the panel below the window is C. So the purple reading is C 78.26.

If there are two lane numbers visible, the centilane reading indicates the correct one.

'Zero' controls. If, in the 'Ref.' setting of the function switch, the three decometer pointers do not indicate zero they can be set to zero with the 'Zero' controls (8). Correction, if any, with these controls should be done particularly:

(*a*) when the receiver has been switched on for less than one hour;

(*b*) when switching from one chain to another;

(*c*) just before an exact position is required.

Suppose the red decometer indicates A 9.40. If the function switch is set to 'Ref.' the pointer should turn along the shortest way to zero or near zero. After correction, if any, the reading will be A 9.00. When the function switch is set back to 'Op.' the pointer resumes its pattern reading. If the ship has moved during this time by 0.2 lane the reading should be A 9.60, but the pointer will go the shortest way (i.e. via 8.90, 8.80 and 8.70) to 8.60 instead of 9.60. To avoid this, the zero adjustment should not take too much time.

The 'Zero' controls are electronically interconnected with the LI section of the receiver, and LI will only operate correctly if the decometers have been correctly referenced.

LI digital readout. LI can be read off on the number tubes (14). This enables the user to adjust each lane indicator on the decometer to the correct lane number for the ship's position when setting up. The LI readout also forms a constant monitor of the satisfactory tracking of the decometers when the ship is under way, as it shows every 20 seconds each position-line value to the nearest tenth of a lane. The readout is scaled 00.0–23.9 for red, 30.0–47.9 for green and 50.0–79.9 for purple.

'LI zero' pushbutton. In normal operation the master LI reading is zero (within the tolerance given later), but on first switching on it will have some other value. The purpose of the 'LI zero' button (7) is to set the master LI reading to zero; once it is pressed and released, subsequent master readings should be zero. The control should be used with caution at night.

LI dimming control. The dimming control (6) adjusts the illumination of the number tubes.

Use of LI when entering the coverage or when setting up within the coverage

The Decca operating instructions and data sheets contain detailed operating instructions for the receiver. The following résumé of two of the operating procedures, taken partly from the manual, is intended to give the reader an impression of the operation and make him conversant with it to some extent. In practice, the Decca instructions should always be consulted.

1. Switch on and set up the Decca Navigator in the prescribed manner.
2. Plot the estimated position of the ship on the Decca chart and from this set the zone letter, lane number and centilanes on each decometer. Naturally these settings will only be as good as the estimate of the ship's position. When the reset knob is released the centilane pointer may move away from the reading you have set by a few tenths of a lane. (The pointer indication is determined by the difference in phase

between the signals of master and slave, and the position of the ship found by this method may differ from the estimated position.) Do not attempt to change this new reading, because the centilane pointer automatically swings along the shortest way to the correct position, producing the lane and centilane reading nearest to the one you have set. For example, if you set the meter to B 17.90 and the true centilane position is 10, the reading will change to B 18.10 (not B 17.10).

3. Observe over several sequences whether the LI sequence of red, green and purple is correct. *Do not proceed with the setting-up procedure until regular and correct sequences are observed.*

4. Observe a group of three complete and, if possible, consecutive colour sequences. Record the LI readings, the decometer readings and (in a third column) the positive or negative difference between them. (In the operating instructions a specimen log is shown as an example). To summarise the remainder of the procedure, the decometers can be considered to be correctly set only when the difference between the LI readings and decometer readings for any colour is consistent and not more than half a lane.

Operation during the voyage

1. During the first hour of operation, and occasionally during each day of operation, the function switch should be turned to 'Ref.' and the centilane pointer adjusted to zero if in error. Switching to and from 'Ref.' can cause the centilane pointers to rotate and thus to lose the lane count; to prevent this, the following procedure is recommended:
   Press the 'Hold' button and keep it depressed.
   Turn the function switch to 'Ref.'
   After 10 seconds, release the 'Hold' button.
   Check the decometer zeros, adjust 'Zero' controls as necessary.
   Press the 'Hold' button and keep it depressed.
   Turn the function switch to 'Lock 2'.
   When 'Lock' lamp is steady, turn the function switch to 'Op.'
   When correct LI is apparent, release 'Hold' button.

2. Similarly, the master LI readings, i.e. the first reading in each LI sequence, should be periodically checked to ensure that they are within the tolerance 23.8 to 00.2. If successive readings are consistently outside this tolerance, press and release the 'LI Zero' button and check the master reading again on the next LI sequence.

Effects of sky-wave interference

At night, and possibly during daylight at long ranges, the LI master readings may fall outside the stated tolerance, but in an inconsistent or random manner. When this effect is observed, do not use the 'LI zero' button and note that all LI readings may be unreliable.

Under these extreme sky-wave conditions, when the master LI readings are within tolerance the fractional (tenths of a lane) red, green and purple LI readings may nevertheless disagree with the decometer centilane pointers. In such cases, the LI reading may provide a more accurate position-line than the corresponding decometer reading.

Lane-slip

It will be appreciated that position-fixing systems can never be absolutely reliable. Therefore the normal check on a Decca fix must never be neglected.

One of the risks with Decca is that lane slip may not be noticed in time. To prevent this one should:

1. Periodically compare the readings of the LI with those of the decometers.
2. Periodically and frequently plot Decca positions and check them with positions obtained from external sources.
3. Listen to radio transmissions of Notices to Mariners, which might contain warnings of incorrect or interrupted transmissions from Decca chains (see data sheets).

Lane-slip can be caused by:

1. Disturbances in the supply voltage (change of dynamo, short-circuiting) or a large deviation from the nominal supply voltage.
2. Simultaneous reception of ground and sky-waves, especially at long distance from the chain.
3. Switching to 'Ref.' when any of the centilane pointers is near 50, or remaining on 'Ref.' for too long a time.
4. Reception of strong atmospheric or man-made interference.
5. Interference by other stations transmitting on Decca frequencies.
6. Damage to the insulation of the aerial, or an inadequate earth-connection on board.
7. Breakdown of, or incorrect transmission from, Decca chains or receivers.

Applications

The Decca Navigator is widely used for general coastal navigation, for fishing, and for harbour-approach navigation. It is also used for position-fixing and survey work. The Decca system can be used on board a ship at anchor in foggy weather to check whether the anchor drags.

Further, the radar screen may happen to show an object, the nature of which is not certain and which may be a buoy, a lightvessel or another ship. As Decca indicates the ship's position accurately, it is possible to determine the position of these unknown objects and identify them.

Decca is particularly useful to cable ships, buoyage vessels, surveying vessels for hydrographic purposes, salvage boats and certain navy craft. The speed and smallest turning circle of a ship on trials are often determined with great accuracy by means of Decca. In 1977 the number of ships equipped with Decca amounted to about 22 000.

Some Decca lines approximately coincide with shipping lanes. In such cases a Decca line can be followed with the help of one of the decometers, keeping the centilane pointer at the same setting. (Remember, however, that since vessels may follow the same Decca line in opposite directions there is an increased risk of collision.) When following a Decca line the possibility of lane slip should be particularly taken into account, and the normal control on the position should not be neglected since an erroneous indication may result in the vessel running aground.

An ancillary item to the Mark 21 is the Remote Decometer Display. This functions as a slave unit to the main receiver and allows the display of decometer-coordinates remotely from the receiver.

THE TRACK PLOTTER

Because of its great accuracy and automatic indication, the Decca system has interesting possibilities for air and maritime navigation, as the position of the craft can be shown automatically and continuously on a chart and the path travelled can be recorded. To achieve this, the rotating movement of the centilane pointer has to be converted into rectilinear movement of a stylus. This stylus would then, if linked to the red decometer, move in the direction perpendicular to the red Decca lines of the chart and simultaneously, if also linked to the green decometer, in the direction perpendicular to the green Decca lines. It is, of course, possible to move the chart instead of the stylus.

Suppose the Decca lines on the chart (e.g. the red and the green ones) are at right-angles to each other. As in *Figure 5.25*, the chart can then be made to move vertically and the stylus horizontally. The first movement is brought about by winding the paper from one cylinder to another; the movement of the stylus is effected by means of a dead-screw. As a rule, however, Decca lines are not at right-angles to each other, indeed their angle of intersection varies, nor are lines of the same colour exactly parallel. However, a chart can be drawn which, although badly distorted, can nevertheless serve a useful purpose.

The red lines, for instance, are plotted on a sheet of paper as parallel straight lines at equal distances from each other. Each line is marked with the letter and number (say B 20). Each strip between two such lines consequently represents, say, a lane. At right-angles to these lines the green

Decca lines are plotted. Then the course line, coastline, etc, are transcribed point by point from the Decca chart to the sheet of paper, with the aid of the Decca coordinates. The chart obtained in this way may, of course, be badly distorted and not always easy to read, but this drawback is counterbalanced for some applications by the advantages of the pictorial display.

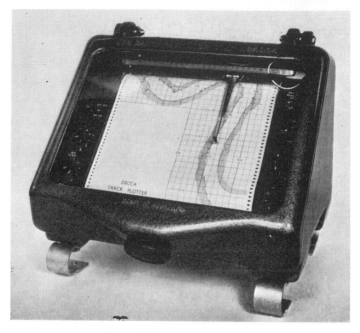

Figure 5.25 Decca track plotter

The principal application of the track plotter is in operations that call for accurate holding of predetermined tracks, or the recording of tracks that can be analysed afterwards. Examples are: fishing, navigation of ferries in congested and hazardous traffic routes, hydrographic surveying and underwater oil exploration, cable laying, buoy maintenance, dredging, sea rescue work and minesweeping. When used in fishing, the ground wrecks and fastenings can be plotted on the track-plotter chart, previous prolific hauls retraced and planned fishing carried out; the helmsman can steer by reference to the pre-drawn course line on the track-plotter chart, and the labour of manual plotting is avoided.

6

The Loran system

Loran is a hyperbolic system of position fixing with a long range. (The word Loran is composed from the words *long range navigation*.) There are two technically different Loran systems, the original Loran-A (still in use) and the improved version, Loran-C. We shall look first at the principle of Loran, and then describe the operation of the two systems in turn.

PRINCIPLE

Loran stations transmit radio signals of very short duration called pulses. The number of pulses per second is called the pulse-recurrence rate or pulse-repetition frequency (p.r.r. or p.r.f.). For the Loran-A system, the p.r.r. amounts to about 20, 25 or 34 depending on the transmitting stations. If the p.r.r. is 25, a pulse with a duration of 40 μs (1 μs = 1 microsecond = 1 millionth of a second) is transmitted every 1/25 second. The

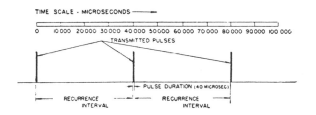

Figure 6.1 Recurrence interval and pulse duration of Loran pulses (courtesy US Navy Department)

recurrence interval (see *Figure 6.1*) is then 1/25 s or 40 000 μs. The transmitter therefore is working for only 40 μs out of every 40 000 μs, or 0.1 per cent of the time. But during this short time the transmitter is very active and the power radiated is high.

As the velocity of radio waves is 300 000 km/s or 300 m/μs a distance

175

can be indicated in terms of the time the radio waves need to travel it; for instance, 9 km or 30 × 300 m corresponds to 30 μs.

Loran lines

In *Figure 6.2*, A and B represent Loran transmitters. The distance between them is 324 nautical miles, corresponding to 2000 μs. Suppose that A and B transmit their pulses simultaneously. If the pulses from A arrive, say, 800 μs later than those from B, we are on the hyperbola CD in *Figure 6.2*.

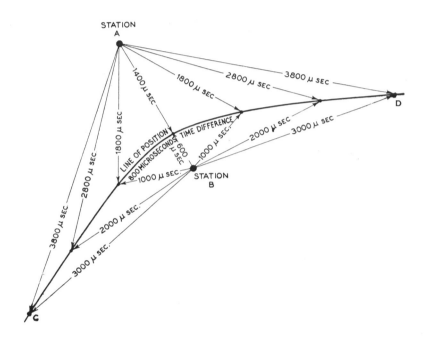

Figure 6.2 Pair of stations with simultaneous pulsing, showing line of constant time difference, CD (courtesy US Navy Department)

Other hyperbolas, for time differences increasing in 200 μs steps, are shown in *Figure 6.3*. Since we cannot distinguish the A pulse from the B pulse, but can only measure the time-difference, we cannot yet say which of two Loran lines, situated symmetrically on either side of the centreline, we are on. If the time difference is 200 μs, for instance, we may be on Loran line R or line S of *Figure 6.3*. This ambiguity might give rise to errors, particularly in the neighbourhood of the centreline.

To avoid this, the following procedure can be adopted. Station A, the master, first transmits a pulse. At the moment when station B, the slave, receives this pulse, B transmits its own pulse. The two pulses will arrive at the same time at a ship on the baseline extension BC (see *Figure 6.4*), whereas on the other baseline extension AD the maximum time difference will be measured, i.e. the time needed by the radiation from A to travel to B plus the time needed by the radiation from B to travel to A. For a ship on the baseline BA the time difference measured increases regularly from zero at B to maximum at A. According to the definition of a Loran line,

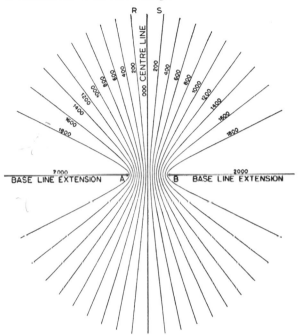

Figure 6.3 Lines of constant time difference from a pair of stations with simultaneous pulsing. For a given time difference there are two possible lines of position (courtesy US Navy Department)

the same time difference occurs at all points of a Loran line; at every point on the hyperbola FG, for instance, the time difference is the same as at E. It follows that a longer time difference is measured on hyperbola R than on hyperbola S in *Figure 6.3*, so the ambiguity has disappeared.

Coding delay

Even now, however, another difficulty may be encountered. Suppose (*Figure 6.4*) that you are at point H in the vicinity of the baseline extension BC, where the pulse from A arrives 30 μs before that from B. Since

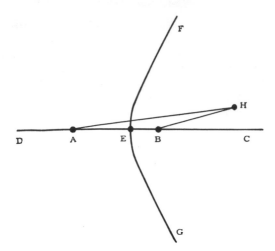

Figure 6.4 If slave B transmits its pulse on receiving the pulse from master A, the two pulses will arrive simultaneously at a receiver on BC; however, they will overlap at H

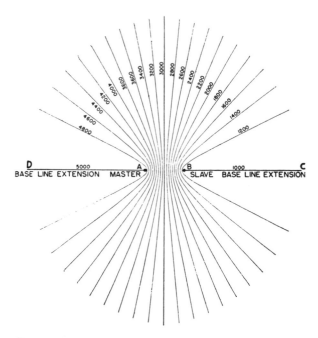

Figure 6.5 Loran lines from actual pair of stations with a coding delay of 1000 μs (courtesy US Navy Department)

the duration of a pulse is about 40 μs, the pulse from B arrives before the pulse from A has ended. The two pulses partly coincide, therefore, and thus measurement of their time difference is not possible. *By postponing the transmission of the pulse from B* for a short time the two pulses will be received separately on board and the time difference can be measured. The time delay between the transmission of the A and the B pulses is then the time that the A pulse takes to go from master to slave, plus the *coding delay*.

The shape of the Loran lines remains with this method the same as in *Figure 6.3* but there are no longer two lines with the same time difference (see *Figure 6.5*). The Loran line with minimum time difference is the base-line extension BC; this time difference, in *Figure 6.5*, is 1000 μs and is the coding delay.

In the slave station, care is taken to ensure that the delay is kept accurately constant, for if this should change, the time difference at a given spot would alter, and the Loran chart would give an incorrect indication of position.

LORAN-A SYSTEM

Measurement of the time difference

With the Loran-A system the pulses are received by an apparatus that is basically similar to an ordinary radio receiver. The signals are not supplied to a loudspeaker or a telephone, however, but to a cathode-ray tube, known sometimes in this case as an *electronic watch*. By means of this apparatus, the difference in time between the arrival of the signal from the master and its arrival from the slave can be measured accurately in microseconds.

A sawtooth voltage is applied to one set of deflecting plates to produce the horizontal timebase (see page 29). The pulses received are applied, after amplification and detection in the receiver, to the other set of deflecting plates, where they cause a vertical deflection. Because of their short duration, the pulse envelopes become visible as small peaks with steep sides (see *Figure 6.6(a)*). The stronger the pulse, the higher the peak.

The sweep frequency is the number of times per second that the timebase is described. If this sweep frequency is made exactly equal to the number of pulses arriving per second from one station (the pulse-repetition rate), the time during which the beam describes the sweep is equal to the recurrence interval, and the successive pulses from one station will always appear on the same spot of the timebase. These pulses are seen then as a fixed image.

If there is a slight difference between the two frequencies, successive pulses appear further to the left or right to the end of the timebase, then reappear at the other end.

If we know the time interval between the arrivals of the A and B pulses, we need only look up in the chart the Loran line for this time (see Chart III). This Loran line is then a 'line of position' (LOP) for our ship. The coding delay is included in the time indicated by the apparatus as well as in the time mentioned on the Loran lines on the chart. Thus it need not be subtracted from the measured time difference.

It will be clear that the distance L in *Figure 6.6(a)* represents the time to be measured. If L were, for instance, one-sixth of the length of the

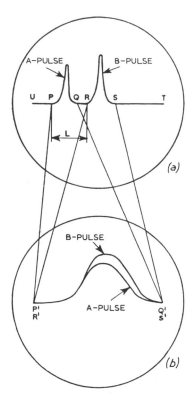

Figure 6.6 Pulses of master and slave are visible on the screen (a) as a static image; L is the time difference to be measured. Use of a fast sweep magnifies the horizontal scale (b), so that PQ and RS become P'Q' and R'S'; when the controls have been adjusted so that the leading edges of the pulses coincide, the digital counter indicates the time difference in microseconds

timebase, the time to be measured would be one-sixth of the time of the recurrence interval. As the latter is exactly known, the former could be calculated.

However, this method is not sufficiently accurate. For the real measuring process only a small part of the timebase, namely the part PQ of *Figure 6.6(a)* on which the A pulse appears, is reproduced enlarged horizontally by a fast sweep. The consequence is that the A pulse is expanded to P'Q' in *Figure 6.6(b)*, and its shape is more clearly visible.

As soon as the spot of light has arrived at point Q' of *Figure 6.6(b)* the electron beam is suppressed for a period that can be adjusted with a coarse-delay control knob and a fine-delay control knob. When after this short period (corresponding to QR in *Figure 6.6(a)*) the B pulse arrives, it is amplified in the same way as the A pulse and reproduced as R'S' in *Figure 6.6(b)*. One can then manipulate the two delay knobs so that the leading edges of the two pulses coincide exactly; this is called *pulse envelope matching* or *pulse matching*. When the light spot, having traced the B pulse in *Figure 6.6(b)*, arrives at the end of the timebase, the electron beam is suppressed again until the next A pulse arrives. The period of suppression corresponds to ST and UP in *Figure 6.6(a)*.

There is a counting device on the apparatus like the mileage-counter of a motor car. As the coarse- and fine-delay controls are turned the digits of this counting device alter. When the leading edges of the pulses coincide, the number of microseconds between the beginnings of both fast sweeps (the time L of *Figure 6.6(a)*) is automatically indicated.

Ground and sky waves

Because the waves travel along different paths, a single transmitted pulse can be received as a series of three or more pulses. The ground wave always

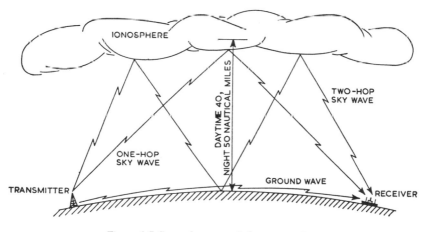

Figure 6.7 Ground-wave and sky-wave paths

travels along the shortest path and so arrives first. Then comes the one-hop sky wave, and afterwards (but not always) other sky waves; see *Figure 6.7*. The beginning of the received ground-wave pulse (the first 35 μs) is free from sky waves, however. With the Loran-A system the range of the ground wave amounts to about 600 nautical miles.

Sky-wave corrections

The time shown on each Loran line on the chart is the difference in arrival of the two *ground* waves. Matching of the ground waves from the two transmitters is to be preferred, even if they are much weaker than the sky waves.

When it is not possible to receive the ground wave from one or both of the transmitters, *the one-hop sky wave from both transmitters should be matched*. As a compensation for the difference in the paths travelled, a correction should be applied to a one-hop measurement in order to convert it to an equivalent measurement of the ground wave, so that the Loran position lines on the chart can still be utilized.

The corrections for a particular pair of stations are shown in the charts and in the tables for various points in their coverage. Between these points one ought to interpolate. The corrections should be added to the measured time difference if they have a plus sign and subtracted from it if they have a minus sign. Since the height of the reflecting layer is liable to alter, the paths travelled and the correction to be applied are not exactly known. The corrections shown are therefore averages, and sky waves must consequently only be utilized if *no ground waves can be received*.

It should be noted here that, on the centreline, the correction is zero if the reflection points for the waves of master and slave are at the same height, which is assumed. Both waves then suffer the same delay.

Identification of the ground and sky waves is very important in Loran position-fixing, and an accurate determination of the position is only possible if the path along which the signals have travelled from the transmitter to the receiver has been ascertained. In 90 per cent of cases, this certainty is obtained if observations are made every hour. The gradually occurring changes can be better followed then, and conclusions can be drawn as to the path travelled by the received waves. In addition, errors will show up by comparison with former positions.

Pulse-repetition rates and radio frequencies

Loran stations do not transmit callsigns; each pair is identified solely by its radio frequency and pulse-repetition rate. Master and slave (situated between 200 and 400 nautical miles apart) use the same radio frequency and transmit the same number of pulses per second; the pulse-repetition rate differs from those of neighbouring pairs.

When a master has more than one slave, the p.r.r. of each master/slave combination differs from that of the others: the master transmits simultaneously two or more series of pulses, each with a different p.r.r. Since all combinations in a chain use the same frequency, a receiver tuned to that frequency will accept the pulses from all the combinations; owing to the difference in p.r.r., however, the pulses of the combination with a p.r.r. not equal to the sweep frequency of the receiver timebase will move across the screen. These pulses always appear at a different point on the timebase, so their brilliance is not enhanced and they are only faintly reproduced; they are called 'ghosts'. In the case of more distant stations the pulses are not visible at all.

There are two controls for the alteration of the sweep frequency and so for selection of the stations, namely:

1. The 'basic pulse-repetition rate' switch with six positions: H, L, S, SH, SL and SS.
2. The 'specific pulse-repetition rate' switch with the positions 0 to 7.

With the first switch we can adjust the receiver for stations with the following basic pulse repetition rates:

H	$33\frac{1}{3}$ pulses/s
L	25 pulses/s
S	20 pulses/s
SH	$16\frac{2}{3}$ pulses/s
SL	$12\frac{1}{2}$ pulses/s
SS	10 pulses/s

The recurrence intervals, in microseconds, for the various positions of the basic and specific p.r.r. switches are shown in the following table.

Specific p.r.r.	Basic p.r.r.					
	SS	SL	SH	S	L	H
0	100 000	80 000	60 000	50 000	40 000	30 000
1	99 900	79 900	59 900	49 900	39 900	29 900
2	99 800	79 800	59 800	49 800	39 800	29 800
3	99 700	79 700	59 700	49 700	39 700	29 700
4	99 600	79 600	59 600	49 600	39 600	29 600
5	99 500	79 500	59 500	49 500	39 500	29 500
6	99 400	79 400	59 400	49 400	39 400	29 400
7	99 300	79 300	59 300	49 300	39 300	29 300

Loran-A charts show, with the Loran lines associated with a particular station-pair, a code consisting of the channel number (1 = 1950 kHz, 2 = 1850 kHz, 3 = 1900 kHz) followed by the basic and specific pulse-repetition rates; e.g. 1SH4.

Tuning

For Loran position-fixing, two station-pairs are needed, each pair providing a position line. The procedure for tuning to each station-pair selected for use is as follows:

1. Select the channel, by setting the channel switch to the number indicated in the code ('1' in the example given above).
2. Select the sweep frequency, by setting the basic p.r.r. and specific p.r.r. switches to the positions indicated by the code ('SH' and '4' in the example).

Pulses from all station-pairs that transmit on the selected channel, and that are not at too great a distance, will then become visible on the screen; but only the station-pair with a p.r.r. equal to the sweep frequency selected will produce a *stationary* image.

Conversely, if a stationary image appears on the screen, the positions of the three switches indicate the station-pair being received.

LORAN-C SYSTEM

Frequency

The ground waves of a *Loran-A* transmitter with a peak power of 1 megawatt (1000 kW) can be received over sea up to a distance of 500–700 nautical miles; at distances up to about 1100 nautical miles ground and sky waves are received. (Over land these distances are considerably shorter.)

Owing to the lower frequency (100 kHz for all transmitters), the ground waves of a 300 kW *Loran-C* transmitter can be received up to 1200 nautical miles away, the one-hop E sky waves up to 2300 nautical miles and the second-hop E even up to 3400 nautical miles. Between 1200 and 1800 nautical miles the ground wave is still sometimes received, but this is unpredictable; accurate observation of the signal in this area is certainly required. At distances of more than 1800 nautical miles, a ground wave can no longer be expected. Unlike Loran-A frequencies, the attenuation and propagation of the 100 kHz Loran-C frequency are practically the same over land as over sea.

For synchronization of the pulse-transmissions of the slave with those of the master, it used to be necessary in both Loran systems for the slave to receive the *ground* waves from the master. The longer ground-wave range of Loran-C over land permitted a longer baseline, and therefore a considerably larger coverage, for Loran-C than for Loran-A. Later on, atomic oscillators became available. Now all Loran transmitters are equipped with them, so neither Loran-A nor Loran-C systems depend on the ground-wave range.

A drawback of the low frequency of 100 kHz is that Loran-C transmitting aerials must have very large dimensions.

Elimination of sky-wave reception

The inaccuracy introduced by reception of sky waves within the ground-wave coverage can be avoided by making use of the fact that these waves arrive later than the ground wave.

As the distance from a station increases, the difference in the distances covered by the ground wave and the sky waves becomes smaller; see *Figure 6.7*. The time between arrival of ground and sky waves therefore decreases as the distance from a station increases; it is about 35 μs at 1000 nautical miles, and not less than 30 μs anywhere in the ground-wave coverage.

The duration of Loran-C pulses is 250 μs (considerably longer than the 40 μs of Loran-A). As a result, the ground wave and sky waves of the same pulse largely overlap each other. However, the Loran-C receiver is now so equipped that on reception of the ground wave normally only the first 35 μs of every pulse is used for measuring the time difference. This part is not yet 'contaminated' by the sky waves. (One may wonder why the Loran-C pulse needs to be 250 μs if only the first 35 μs are used by the receiver. It can be proved, however, that there is less interference produced on neighbouring frequencies when the pulse is made longer.)

The ground wave is attenuated more than the sky wave, and at distances greater than 1000 nautical miles the sky wave is stronger than the ground wave. Outside the ground-wave coverage the sky wave therefore has to be used; owing to its longer path, however, its travelling time differs from that of the ground wave, and when relying on the sky wave one must apply corrections and accept a lower accuracy. The travelling times of sky waves are reasonably stable, except for about half an hour at sunrise and sunset.

The possibility, with *pulse* systems like Loran, of eliminating the unwanted sky-wave mode of propagation is an important advantage over *continuous-wave* systems like Decca, where this is not possible.

Cycle matching

As we have seen, with Loran-A the received pulses are amplified, detected, amplified once more and finally made visible on the screen. By matching the pulse envelopes of master and slave on the screen (*Figure 6.6*) the time-difference, and thereby the difference in distance from the two stations, can be determined.

The Loran-C receiver uses not only pulse-envelope matching but also a more accurate method, *cycle matching*. This measures the difference in phase between the radio frequency of the master signals and that of the slave signals (as in the Decca system, except that Loran signals are pulses, not continuous waves).

For this measurement the Loran-C receiver is equipped with a flywheel oscillator, which acquires the frequency (100 kHz) and phase of the incoming master pulses. In this way the receiver is provided with a continuous reference frequency that has the same phase as the master signals and that retains this phase during the intervals between pulses.

It should be remarked in this connection that, in order to transmit pulses of constant frequency, Loran stations are equipped with oscillators that set up *continuous* oscillations of the correct frequency. Oscillations from this source are then supplied to the transmitting aerial only during the required pulse periods. In connection with cycle matching there must also be phase-synchronization between master and slave aerial currents.

The difference in phase between the frequency of the flywheel-oscillator and that of the slave pulses arriving after those of the master (both 100 kHz) is measured in the receiver. Since one cycle at this frequency lasts $1/100\,000$ s or 10 μs, and a difference in phase can be measured up to about $1/100$ cycle, the accuracy of this process amounts to about $1/100 \times 10 = 0.1$ μs. For example, suppose that a phase difference of 0.35 cycle is measured; since one cycle lasts $10\,\mu$s, 0.35 cycle lasts $0.35 \times 10 = 3.5$ μs.

An analogy of the combination, in Loran, of coarse and fine measurement is two mechanical clocks, one of which is very accurate and has only a seconds hand; the other, less accurate, has only a minute hand. If the former indicates 24 seconds and the latter between 18 and 19 minutes (say 18½ minutes), the time is 18 minutes and 24 seconds, not 17.24 or 19.24.

In the Loran-C receiver the pulse-matching method determines the tens, hundreds, thousands, etc., of microseconds of time difference; the cycle-matching method determines the units and tenths of a unit. If, for instance, the coarse method shows a time difference of 15 886 μs and the finer cycle-matching method 8.1 μs, the time difference is $15\,880 + 8.1 = 15\,888.1$ μs, not 15 878.1 μs or 15 898.1 μs. The accuracy of the coarse method (about 4 μs) is sufficient to justify the rejection of the latter two.

Transmission of Loran-C pulses: time sharing

A series of *eight* pulses is transmitted every recurrence interval with Loran-C, instead of *one* as with Loran-A. The series is first transmitted by the master, some time later by a slave, and thereafter possibly by other slaves (see *Figure 6.8*).

When transmissions are not correct, the users must be warned. The master therefore transmits a ninth pulse: on Loran-C receivers equipped with a cathode-ray tube this ninth pulse will flicker (so-called 'blinking') when a fault in transmission occurs; should the master fail, the pulses of the slave(s) will move to and fro on receiver screens.

An advantage of transmitting eight pulses instead of one is that the influence of noise and other interference on a single pulse, i.e. a mutilation of it, is more or less compensated, from a statistical point of view, by opposite mutilations of other pulses from the series of eight.

The coding delay between the series of eight master pulses and the next series of eight slave pulses is sufficiently long to prevent pulses of the stations of a chain from arriving simultaneously anywhere in the coverage.

Figure 6.8 Time-sharing of master pulses with pulses from two slaves; for an explanation of the ninth master pulse, see text

The time difference between the arrival of each of the eight pulses of the master and the corresponding pulses of a particular slave is measured by the coarse and fine methods. A counter in the receiver shows this time difference in microseconds and tenths of a microsecond. The digits just before and after the decimal point are obtained with the cycle matching method, the higher-significance digits with the pulse-envelope matching method.

Loran receivers can be equipped with one or two counters. If there are two, one of them indicates the time difference between the pulses of master and, say, slave X, and the other the time difference between master and, say, slave Y. The receiver thus determines simultaneously two LOPs, and it is therefore not necessary to apply a 'running fix'.

If the distance from the transmitting stations is too long to receive the ground wave the receiver reacts to the sky wave.

Phase coding of pulses

The oscillations of some of the eight pulses of each station are 180° out of phase with those of the other pulses. This *phase coding* assists, among other things, in the process of initial locking-in, particularly in automatic receivers. The code, which is the same for all Loran chains, is shown below. A plus stands for a pulse that is in phase, a minus for one that is 180° out of phase.

	Master	Slave X	Slave Y	Slave Z
cycle 1	+——+++++	+—+—++ ——	+—+—++ ——	+—+—++ ——
cycle 2	++——+—+—	+++++——+	+++++——+	+++++——+

The third and fourth cycles are the same as the first and second cycles, etc.

AUTOMATIC LORAN-C RECEIVERS

Some Loran-C receivers automatically apply pulse-envelope matching and cycle matching. (With other receivers it is only possible to apply pulse envelope matching manually, with the aid of a cathode-ray tube, as described earlier.) The high price of automatic receivers, formerly an objection to their use, has been considerably reduced for some years.

Automatic pulse-envelope matching

The pulses received from the *master* are amplified and detected; the resulting pulses have the form of the *envelope* of the incoming pulse (see *Figure 1.11*). The automatic Loran-C receiver incorporates a pulse generator, whose pulses are made to start at the same time as the arriving master pulses. (Basically, the method used is that of the flywheel oscillator described in Chapter 1, except that pulses rather than sine-wave oscillations are involved.)

The pulses set up by the generator are supplied to a variable-delay unit. The pulses appearing at the output of the delay system are automatically made to coincide with the pulses arriving from the slave (remember that *everywhere* in the coverage the slave pulses arrive later than the master pulses). The flywheel-oscillator principle is again used: as long as the starts of the delayed pulses and the slave pulses do not coincide, an error voltage is produced that adjusts the delay in such a way that the starts of the two sets of pulses are brought together and the error voltage itself reduces to zero.

The delay being applied to the internally-generated pulses is now the same as that between the received master and slave pulses. It is displayed on a digital counter in tens, hundreds, thousands and tens of thousands of microseconds. The automatic pulse-envelope matching process having thus measured the coarse delay, fine-delay measurement is carried out by automatic cycle matching.

Automatic cycle matching

The phase of the first cycle of the master pulse cannot be compared with the first cycle of the slave because these cycles are too weak and the sampling point is not sufficiently defined. The received pulse is therefore amplified and the pulse obtained (oscillation 1 in *Figure 6.9(a)*) again amplified 1.35 times and shifted 180° in phase. Oscillation 2 thus obtained is algebraically added to the original oscillation 1, to produce oscillation 3 as shown in *Figure 6.9(b)*; this has a well-defined minimum at the sampling point S, which occurs *exactly 30 µs after the start of the received pulse.*

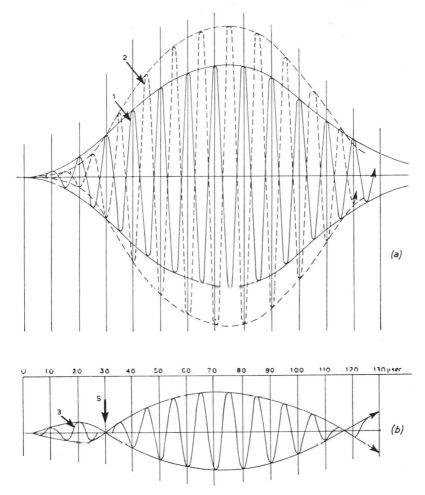

Figure 6.9 Loran-C cycle matching. (a) Oscillation 1 is the amplified pulse with its early part uncontaminated by sky waves; oscillation 2 is obtained by amplifying 1 and shifting the phase 180°. (b) Oscillation 3 is the algebraic addition of oscillations 1 and 2; it has a clearly defined point S suitable for measurement purposes

At this point the third cycle of oscillation 1 is at its peak. The difference in time (in microseconds and tenths of microseconds) between the sampling point S of the master pulse and that of the slave pulse is determined by the same process as for automatic pulse-envelope matching, and shown on the counter. (Note that the method of automatic cycle matching outlined above does not apply to the Loran-C receiver DL 91 described at the end of this chapter.)

The combined result of the coarse- and fine-delay readouts is the time-difference in microseconds, and refers to an LOP on the chart. The same automatic process takes place simultaneously for the other slave.

Automatic tracking continually updates its results at least five times per second. Owing to the continuous indication of two LOPs, the data obtained can be passed on to auxiliary apparatus for automatic processing. Thus latitude and longitude, the course to be steered, etc. can be calculated. The position can also be plotted automatically.

LORAN-A AND LORAN-C CHARTS AND TABLES

Loran-A and Loran-C charts published by the American Oceanographic Office are Mercator charts overprinted with Loran lines. Nautical data such as depths, lighthouses, lightvessels, etc. are not indicated. They serve, therefore, only as position-construction charts.

Depending on the scale of the chart, the time-differences of successive Loran-C lines increase by 20, 50, 100 or 200 μs, so as a rule we have to interpolate when plotting from the receiver readout. The LOP should always be plotted from the nearest Loran line shown on the chart. Near the baseline extensions proportional interpolation is not admissible; this is evident from Chart III.

The station-identification codes printed on Loran-A charts are described on page 183. On Loran-C charts the stations are indicated by: M = master; W, X, Y and Z = slaves. The significance of, for example, S3-X-13 300 printed near a Loran line (Chart III) is: S = basic pulse rate of 20; 3 = specific pulse rate; X = station-pair is master and slave X; 13 300 = time difference for this position line is 13 300 μs.

On Loran-C charts, sky-wave corrections are preceded by SG or GS, where S means the sky wave and the G the ground wave. The first of the two letters SG or GS refers to the mode of reception from the master, the second to that from the slave. An example is:

SS0.W	SG + 47D
SS0.W	SG + 70N
SS0.X + 09D	
SS0.X + 16N	

The first two lines refer to readouts, in the area concerned, for the station-pair master/slave W (pulse rate SS0); by day (D) the readout should be corrected by + 47 μs if the sky wave is being received from the master and the ground wave from slave W; by night (N) the correction should be +70 μs. Evidently in this area we can rely on receiving the ground wave from the slave but, under normal conditions, only the sky wave from the master.

The third and fourth lines refer to readouts for the station-pair master/ slave X: a correction of +9 μs should be applied by day, +16 μs by night. Corrections are indicated near points of intersection of whole degrees of longitude and latitude. When a correction has decreased to zero, it is not indicated any further.

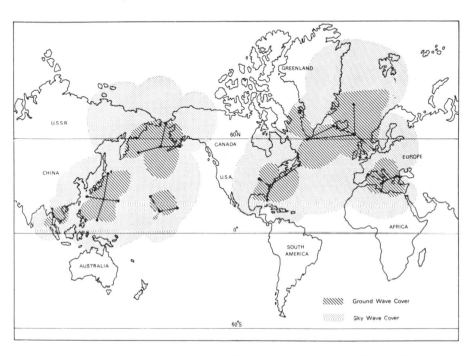

Figure 6.10 Ground- and sky-wave coverages of Loran-C chains (courtesy US Navy Department)

Nowadays there are 65 Loran-A chains and 8 Loran-C chains, principally in the Northern Hemisphere (Atlantic and Pacific Ocean, Mediterranean, etc.). *Figure 6.10* is a map of the ground- and sky-wave coverages of the Loran-C chains existing in 1977.

DESCRIPTION OF A LORAN-C RECEIVER

Figure 6.11 shows the automatic Loran-C receiver DL 91 of the Decca Navigator Co. The following description is largely summarised from the Operator's Manual. All controls are simple and self-evident; those normally not used are concealed behind a flip-open cover along the lower edge of the front panel.

Modes of operation

There are two possible modes of operation, for use under different reception conditions.

The *track mode* of operation is used in areas where good ground-wave coverage is available, i.e. in locations within approximately 1000 nautical miles of the most distant transmitter of the chain being used. In this mode of operation, a constant rechecking of third-cycle synchronization is

Figure 6.11 Automatic Loran-C receiver type DL91 (courtesy Decca Navigator Co. Ltd)

carried on within the receiver, and maximum accuracy and confidence level are obtained. The track mode may be continued as long as the 'Signal' indicators are illuminated and all the 'Sky/Syn' indicators (*Figure 6.12*) are extinguished.

As the receiver moves further away from the transmitters, conditions will arise in which received signals become attenuated to the extent that they are no longer usable for operation on the third cycle. This will usually occur at ranges in excess of 1000 nautical miles, and is evidenced by one or more of the green 'Signal' indicators flickering or extinguishing. Under these conditions, the range of operation will be extended by switching to the *extended-range mode*. In this mode, the synchronization point is shifted automatically from the *third* to the *seventh* cycle of the received Loran-C pulse; as can be seen from *Figure 6.9(b)*, the seventh is the strongest cycle. Reception of sky waves at extended ranges reduces the accuracy of this mode, however, compared to that of the track mode.

As a consequence of pulse transmission, Loran-C operates in a very broad frequency band (about 70–130 kHz). To reduce interfering signals the receiver bandwidth could be made smaller, but this would influence the shape of the pulses and consequently the pulse matching. Provision is made for rejecting unwanted signals that might be detrimental to the correct operation of the receiver. Four fixed rejection filters can be incorporated into the antenna coupler as an optional feature. These would be preset to eliminate any interference signals known to be constantly present in any given area(s) in which the receiver is to be used. Two further filters, tunable over the range 70 to 130 kHz, are incorporated into the receiver itself; these filters, must not, however, be tuned between 90 and 110 kHz (the '100 sector' mentioned below).

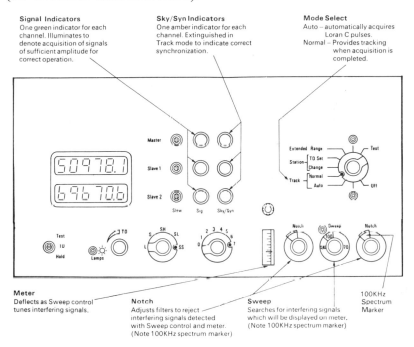

Signal Indicators
One green indicator for each channel. Illuminates to denote acquisition of signals of sufficient amplitude for correct operation.

Sky/Syn Indicators
One amber indicator for each channel. Extinguished in Track mode to indicate correct synchronization.

Mode Select
Auto – automatically acquires Loran C pulses.
Normal – Provides tracking when acquisition is completed.

Meter
Deflects as Sweep control tunes interfering signals.

Notch
Adjusts filters to reject interfering signals detected with Sweep control and meter. (Note 100KHz spectrum marker)

Sweep
Searches for interfering signals which will be displayed on meter. (Note 100KHz spectrum marker)

100KHz Spectrum Marker

Figure 6.12 Front panel of Loran-C receiver DL91 (courtesy Decca Navigator Co. Ltd)

As a Loran receiver has no telephone, interfering signals are detected by means of the 'Sweep' control and the panel meter. Once these signals have been detected, the rejection filters may be adjusted by means of the two 'Notch' controls as detailed in the procedure given below. The 'Sweep' and 'Notch' controls are clearly marked with the spectrum (90–110 kHz) of the received signals. The 'Notch' controls *must not* be left within the limits of this spectrum (the '100 sector'), since this would cause the Loran-C signals themselves to be rejected.

Track mode operation

The controls and indicators used in the track mode are illustrated in *Figure 6.12*. For *initial acquisition and tracking*, use the following procedure:

1. Determine from Loran-C charts which Loran chain is appropriate for the area. Set the two chain select switches to the pulse-rate code for this chain.
2. Push in and rotate the mode select switch from 'Off' to 'Auto'.
3. Depress the 'Sweep' control and tune to the highest meter peak *outside* the '100 sector'. Adjust one of the 'Notch' controls to dip the meter reading. Repeat this procedure with the 'Sweep' control, and notch out the next highest meter peak with the second 'Notch' control.
4. Within 1½ minutes, all 'Signal' indicators should be illuminated, indicating that initial acquisition is complete. Within 10 minutes, all 'Sky/Syn' indicators will be extinguished, indicating that third-cycle synchronization has been accomplished. At this time, set the mode select switch from 'Auto' to 'Normal'.

The receiver will now continue to track and update time-difference readings automatically. The foregoing procedure assumes operation in good ground-wave coverage during daylight hours.

Extended-range operation

When the limits of ground-wave operation are reached (1000–1100 nautical miles all-sea path), extended-range operation may be used to obtain further coverage. The need for extended-range operation is indicated by one or more of the 'Signal' indicators flickering or extinguishing. As soon as this occurs, the mode select switch must be set from 'Normal' to 'Extended Range'. This causes the sampling point within the receiver to shift automatically to the seventh cycle, and all the 'Sky/Syn' indicators to illuminate, warning the operator that reduced accuracy may be expected.

7

The Omega system

In 1940 the Decca Navigator Co. was already studying the possibility of a
hyperbolic system of position fixing with a very long range, operating on
the same principles as the present Decca system. The frequency would be
10.2 kHz, the same frequency at which the Omega system now operates.
In the same year Decca intended to build stations for this system in
Ireland. The Irish Post Office authorities feared, however, that the very
low frequency of 10.2 kHz (which, when converted to sound waves, is an
audible frequency) would induce interference signals in telephone lines. So
the plans were not realised. A second proposal for a hyperbolic system in
the very low frequency band, the Delrac system, was not realised either.

In 1957 the US Navy started development of the Omega system, operating on the same principle as Delrac. Alpha and omega are the first and last
letters of the Greek alphabet, and so may be used to symbolise the begin
ning and end of a development. Presumably this has led to the name
'Omega'.

The system in its present form, with seven transmitting stations, covers
practically the entire world; with the proposed eight stations, the Omega
system will indeed cover the whole globe. Not only ships but also aircraft
and submerged submarines (at a maximum depth of about 10 metres) can
use the system.

Propagation of radio waves at very low frequencies

The weak point of hyperbolic position-fixing systems with regard to the
improvement of accuracy always lies in anomalies of radio-wave propagation, which depend on circumstances that cannot be controlled.

With the exception of frequencies higher than about 40 MHz, radio
waves are reflected by layers of ionized air. The heights of these layers,
and their degree of ionization, are subject to variations. As a result, the
propagation time of the waves between transmitter and receiver varies.

This results in phase variations of the received waves, and ultimately decreases the accuracy of the system in which they are used. (The Earth's magnetic field also influences the propagation of the waves used by Omega.)

Radio waves in the very low frequency (v.l.f.) band, i.e. those between 10 and 30 kHz, are reflected by the *D-layer* of the ionosphere – unlike high-frequency (3 to 30 MHz) and medium-frequency (0.3 to 3 MHz) waves, which are reflected by other layers. The characteristics (especially the accuracy) of a long-range radio position-fixing system are therefore closely related to the frequency at which the system operates.

The Omega system uses v.l.f. radio waves of 10.2 kHz frequency (about 30 km wavelength), which are reflected between the D-layer (at a height of about 80 km) and the surface of the Earth. The loss of power during their propagation and reflection (the attenuation) is low; thus the range can be very long and can be far more than half the circumference of the Earth.

Figure 7.1 indicates the position and joint coverage of the four Omega transmitting stations as they existed in 1971. The station close to New

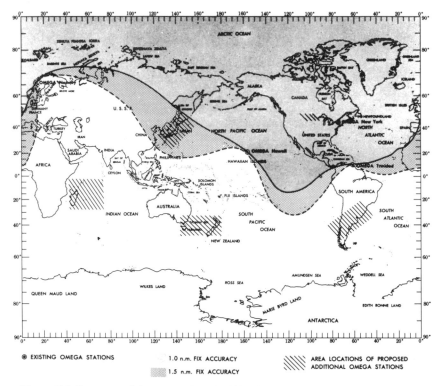

Figure 7.1 Coverage of Omega stations in 1971 (courtesy US Naval Oceanographic Office)

York has since been moved to La Maure, North Dakota. In late 1977, the station in Trinidad having been decommissioned, the operational stations were: Norway, North Dakota, Japan, Hawaii, Liberia, Argentina and Réunion.

Synchronization of transmitter aerial currents

The Omega (like the Decca) system is a hyperbolic method of position fixing based on measuring the phase difference between continuous-wave signals from pairs of transmitters. The phase difference between the alternating currents in the transmitting aerials of a pair of stations need not be zero but must be constant, otherwise the measurements made at the receiver will not accurately indicate the hyperbola on which the receiver is located.

As described in Chapter 5, a Decca slave, on receiving the signals transmitted by the master, amplifies them and supplies them to its transmitting aerial; thus the frequency and phase of master and slave signals are synchronized. The signals received by the slave should preferably be ground waves, in order to avoid the anomalies of propagation time (or 'transit time') inherent in sky-wave propagation, and the distance at which ground-wave reception can be guaranteed is not very great. Hence the distance between master and slave (the baseline) is restricted (in the Decca system the baselines are about 100 nautical miles).

For the Omega system (and nowadays also more and more for other hyperbolic systems) the aerial currents need not, in principle, be synchronized in this way. This is because every station is provided with one or more caesium atomic oscillators. Such an oscillator generates oscillations whose frequency, compared with that of any other kind of oscillator, is extremely stable. The atomic-oscillator frequency is converted to 10.2 kHz and, after amplification, is supplied to the transmitting aerial.

Equipping each Omega transmitter with such an atomic oscillator ensures that the difference in phase between the aerial currents remains constant for a long time. The deviation of the oscillator frequency is only a few cycles in 10^{12}. Nevertheless the slightest deviation in the phase of a transmitting aerial current is corrected.

$\times$ The synchronization of Omega transmissions should meet two requirements. In the first place the aerial alternating currents of all Omega transmitters must maintain a constant phase relationship; this is fundamental to the system's operation. Secondly the frequency must be in accordance with an international timebase, for which the universal time UT2 is taken; this is necessary for the use of Omega transmissions as an international standard of time.

In order to obtain a still better frequency stability there are three atomic oscillators at every Omega station, the average frequency and phase of the three being taken.

Advantage of long baselines

The use of atomic oscillators allows Omega transmitters to be located, without synchronization difficulties, at considerable distances from each other. Their baselines can have a great-circle length of 5000 nautical miles or more, so the system needs very few transmitters for worldwide coverage.

Apart from slight corrections to the phase of its aerial current, each Omega transmission station is independent of the others, so we cannot speak of master and slave. To determine a locus or 'line of position' (LOP) of a ship, any Omega transmitter can be paired with any other Omega transmitter. (In the Decca and Loran systems a master can only be paired with one of its own slaves, although theoretically two slaves of the same master could be used to determine an LOP.) Hence the Omega system is not divided into chains, or rather all Omega stations together form one chain. In the North Sea, for instance, it is possible to receive four stations and to use all four for position fixing.

With four stations A, B, C and D, six combinations are possible: A-B, A-C, A-D, B-C, B-D and C-D; with eight stations there are 28 combinations. (Note that the letters of a combination are always given in alphabetical order; see page 206).

In *Figure 1.42* it was shown that a long baseline considerably increases the area of high accuracy. The ratio of the baselines in that figure was only 1:3. However, an Omega baseline is roughly fifty times as long as, for instance, a Decca baseline, so the widening or expansion of Omega lanes away from the baseline is considerably less than *Figure 1.42* suggests. The small widening of Omega lanes is a great advantage over other hyperbolic systems. Suppose, for instance, that the phase difference can be measured to an accuracy of four centilanes; as the lane-width increases away from the baseline, the error expressed as a *distance* increases. Therefore the less the increase in lane-width the greater the accuracy.

Transmitting aerials

Long-range v.l.f. transmissions need very long aerials. The length of a transmission aerial for such waves should theoretically be a quarter of a wavelength. At 10 kHz the wavelength is 30 km, and a quarter wavelength is 7.5 km. It is true that, by inserting coils in the aerial circuit, we can manage with shorter wires, but even then very long wires and high aerial masts are required for 10.2 kHz. (Unlike these very long *transmitting* aerials, a length of 3 to 20 metres is sufficient for an Omega *receiving* aerial.)

Figure 7.2 is a drawing of an umbrella transmitting aerial. *Figure 7.3* shows the valley-span aerial at Aldra, Norway. It can be seen from the map that the aerial-wires are 4.4 km and 5 km long. Making use of the local topography, the stations Hawaii and Réunion have similar aerials.

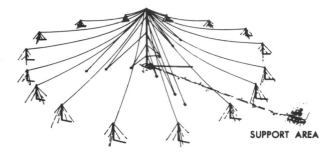

Figure 7.2 Umbrella aerial of an Omega station (courtesy US Naval Oceanographic Office)

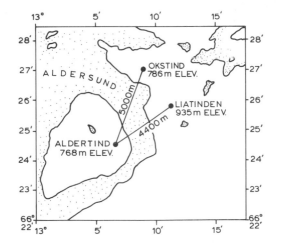

Figure 7.3 The valley-span aerial of the Omega station at Aldra, Norway (courtesy US Naval Oceanographic Office)

Because of these long aerials, and for other reasons, the transmitting stations are expensive. The US has invested great amounts in the entire Omega project. Because of the large coverage the cost per unit of coverage is, however, less than that of any other hyperbolic system.

Waveguide effect

As mentioned before, the space in which radio waves of about 30 km are propagated is bounded by the surface of the Earth and by the D-layer, whose height by day is about 75 km and by night about 88 km. The propagation within this space shows similar characteristics to that in a radar waveguide, in which there are different modes of propagation. (It is true that in a waveguide the height is only a few centimetres while with Omega it is about 80 km, but the respective wavelengths are in proportion: about 3 cm in radar, 30 km in Omega.)

With the Omega frequency of 10.2 kHz there are two modes, one predominating over the other. The second-mode radio waves suffer more loss of energy than the first-mode waves, so the 'waveguide effect' only plays a part up to about 650 n. miles. The error arising from the effect may be as much as two nautical miles. As a warning against this extra error the LOPs on the Omega chart as shown dashed within 650 n. miles of a transmitter. The use of an Omega station for position fixing should be avoided if its LOPs on the chart are dashed.

Polar cap absorption

A second (and originally unforeseen) effect of transmission at very low frequencies is called *polar cap absorption* (PCA). Radio waves arriving via arctic regions such as Greenland and parts of Iceland show rapid phase changes. It was discovered that the cause is not the ice cap but the focusing effect of the Earth's magnetic field, which attracts particles released from the sun and concentrates them in the region of a pole of the Earth. These particles influence the phase and as a consequence the fix accuracy. PCA plays a particularly large part at times of increased solar activity. Because of PCA it is recommended that one should not make use of radio paths via the polar cap.

The Earth's magnetic field gives rise to another effect (fortunately a less harmful one), namely that transmitters have a longer range in the western than in the eastern magnetic direction; with the very long range of Omega transmitters this effect becomes noticeable.

The variations in the propagation of v.l.f. radio waves depends therefore on various circumstances. The phase variations measured at a given position, however, are approximately the same from day to day, as shown in *Figure 7.4.*

Time sharing

A hyperbolic system operating on continuous waves is based on measurement of the phase difference between equal frequencies received from two different stations. If these frequencies arrive simultaneously, however, they will merge into one frequency and the information about the phase difference will be lost.

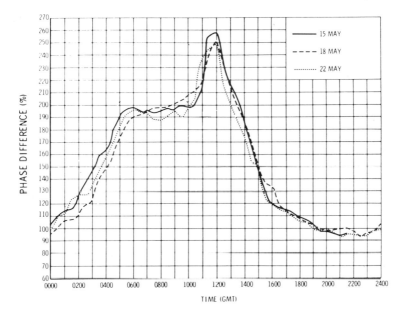

Figure 7.4 Phase-difference variations on three non-consecutive days in May 1972 in Washington DC (courtesy US Naval Hydrographic Office)

In order to preserve this phase information, in the Decca system (as explained in Chapter 5) the equal master and slave frequencies are converted, before transmission, to two different frequencies so that selective circuits in the receiver can keep them separate. The two frequencies are then reconverted to identical frequencies and their phase difference is measured.

Omega uses a different method, 'time sharing'. *Figure 7.5* shows how this will be operated with the proposed eight stations. Station A operates for 0.9 s on 10.2 kHz, then (after a short interruption of 0.2 s) for 1.0 s on 13.6 kHz, and (after another 0.2 s interruption) for 1.1 s on $11\frac{1}{3}$ kHz. The transmissions are organised so that, for example, when D is operating on 10.2 kHz, C is on 13.6 kHz and B on $11\frac{1}{3}$ kHz. The phase during each transmission of about one second is of course the same as that of the previous transmission at the same frequency from the same station; the transmitter oscillator generates a continuous signal of constant phase, but its transmission at each frequency is interrupted during about nine of the 10 seconds.

Omega stations are very modest users of the frequency spectrum. The eight stations will use only three frequencies, and there will never be more than one station transmitting on each frequency at a given moment. In *Figure 7.5* the frequency f_1 from A, f_2 from B, etc., can be used for special purposes. This subject, and also the use of the frequencies 13.6 and $11\frac{1}{3}$ kHz, is considered later.

TRANSMISSION INTERVAL	0.9	1.0	1.1	1.2	1.1	0.9	1.2	1.0	0.9
STATION A	10.2	13.6	11 1/3	—f_1→					10.2
B	—f_2→	10.2	13.6	11 1/3	—f_2—				
C		—f_3→	10.2	13.6	11 1/3	—f_3—			
D			—f_4→	10.2	13.6	11 1/3	—f_4—		
E				—f_5→	10.2	13.6	11 1/3	—f_5→	
F					—f_6→	10.2	13.6	11 1/3	—f_6—
G	11 1/3					—f_7→	10.2	13.6	11 1/3
H	13.6	11 1/3					—f_8→	10.2	13.6

START ← 10 SECONDS → START · ETC.   —►|←— 0.2 SEC.

Figure 7.5 Transmission format of time-sharing Omega stations. The frequencies f_1 to f_8 can be used for special purposes

The phase difference between two signals that arrive at different times can be measured in the receiver by using a flywheel oscillator (see Chapter 1). Such an oscillator acquires the frequency and phase of a signal supplied from outside, in this case a signal of about one second's duration from one of the two Omega stations chosen to provide an LOP. The stability of the flywheel oscillator is such that it retains practically the same phase and frequency until the next one-second signal at the same frequency from the same station, about nine seconds later.

If the ship has moved away from, or towards, the transmitter in those nine seconds, the phase of the new signal will be very slightly different from that of the previous one. The flywheel oscillator responds to this phase change, and retains the new phase for the next nine seconds.

A second flywheel oscillator similarly acquires the phase and frequency of signals from the other Omega station, and responds to phase changes resulting from the ship's movement relative to that station. A discriminator can then continuously measure the phase difference between the oscillator outputs. This indicates the ship's LOP within the lane. Modern receivers show the result directly in digital form.

If a ship crosses one lane the phase difference changes by 360° or one cycle, divided into 100 equal *centicycles* (abbreviated to cec). A phase-difference change of one centicycle arises from the movement of the ship across 0.01 lane or one *centilane* (abbreviated to cel). Note that centilane width varies within the lane (for the same reason that, in *Figure 1.41*, FG < GH < HI, etc.). These differences in width are negligible, however.

Lane identification

As described so far, the Omega system makes it possible to ascertain an LOP *within a lane* but not to ascertain which lane the ship is in. Now, the receiver remains tuned in permanently, even in port; once the lane indication is correct it should theoretically remain correct. Experience has taught, however, that lane slip sometimes occurs. Omega is often used for ocean navigation, in which case lane slip may remain unnoticed or, even if it is observed, the lane may not be ascertainable from other sources.

The minimum lane width (at the baseline) is about 8 n. miles. Hence, to be sure which lane the ship is in, it would be necessary to determine the position by another method of position fixing with an accuracy of at least half of this, i.e. 4 n. miles. This is not always possible on the ocean, so the Omega system has to be equipped with a means of *lane identification* (LI).

In principle, LI in the Omega system operates in the same way as in the Decca system: the frequency on which the phase difference is measured (the comparison frequency) is temporarily lowered, producing a correspondingly wider lane. The receiver finds, by the usual phase-difference method, the ship's LOP in the *wide* lane, and hence finds which *normal* lane the ship is in.

The lower frequency necessary for LI is obtained, in the Omega system, by using both the 10.2 kHz and the 13.6 kHz signals received from each station (see *Figure 7.5*). By mixing the two frequencies, the difference frequency 13.6 − 10.2 = 3.4 kHz is obtained. This is done for each of the stations providing the LOP. The phase difference between the two 3.4 kHz signals is then measured, and indicates the ship's LOP in the wide lane.

The only difference between LI and normal reception is that the comparison frequency has been reduced from 10.2 to 3.4 kHz. Now, 3.4 kHz is exactly one third of 10.2 kHz. Since wavelength is inversely proportional to frequency, a frequency three times as low makes the wavelength three times as long. The width of a lane, as measured at the baseline, is equal to half a wavelength, so a wide lane of the 3.4 kHz pattern is equal to three lanes of the 10.2 kHz pattern. On the baseline this is 3 × about 8 = about 24 nautical miles.

As the 10.2 kHz and the 13.6 kHz originated, at each transmitter, in the same atomic oscillator, the limits of such a wide lane coincide exactly with the limits of three normal lanes. Therefore it is not necessary to indicate the wide lanes on the chart separately.

One might expect that, as in the Decca system, there would be a separate LI display. This is not the case in the Omega system, however. If one switches over to LI (3.4 kHz pattern) the result is given on the same display as before, and in the same units (lanes and centilanes) as before.

It is not recommended to switch a ship's receiver continuously to LI, as aircraft sometimes do, since the accuracy of LI in nautical miles is less than that of the normal 10.2 kHz pattern. Take the case, for instance, of a lane width of 10 n. miles. A wide lane is then 30 n. miles. An accuracy of, say, four centicycles corresponds to 0.04 $\times$ 30 = 1.2 n. miles when the apparatus is switched to LI. For the normal 10.2 kHz pattern the accuracy is 0.04 $\times$ 10 = 0.4 n. mile, considerably better.

When switching back from LI to 10.2 kHz, the readout in lanes and centilanes should theoretically remain the same, but in practice is likely to be slightly different. If the *lane* number changes, it should, in principle, be reset to the LI value, taking sky-wave corrections (page 205) and other sources of position-fixing into account; in general with LI the *lane* number is more reliable, but with the normal 10.2 kHz frequency the *centilane* number is more accurate. However, *the LI indication should not be relied on implicitly*. If the LI readout is, for instance, 693.95, the number of centilanes given in the normal 10.2 kHz readout finally decides whether the ship is in lane 693 or 694: if the number of centilanes in the 10.2 kHz readout is 10, the correct readout is 694.10 (and not 693.10).

It may occur that lane slip is so great that we do not know with certainty in which wide lane of the 3.4 kHz pattern the ship is located. (This is sometimes a problem for aircraft with their high speed.) In that case we can obtain still wider lanes by using the $11\frac{1}{3}$ kHz signals (see *Figure 7.5*). Mixing $11\frac{1}{3}$ kHz with 10.2 kHz produces a frequency of $11\frac{1}{3} - 10\frac{2}{10} = 1\frac{2}{15}$ kHz; this is exactly nine times as low as 10.2 kHz, so the new LI lanes are nine times as wide (and three times as wide as the previous LI lanes). Their width, measured on the baseline, therefore amounts to 9 $\times$ 8 = 72 n. miles. It is possible that, for the purpose of air navigation, still wider lanes will be available in future.

See page 215 for the method of checking lane number in port without the use of LI.

Diurnal propagation variations

The lengths of the paths followed by radio waves between the two Omega transmitters and the receiver (and hence the phase difference) depend on the height of the reflecting layer and the extent of its ionisation. The height of the D-layer, important for Omega, shows diurnal changes; it is about 75 km by day, 88 km by night. *Diurnal shift* is the term applied to the change in phase difference brought about by a change in the height of the ionosphere from day to night.

Because of the influence of sunlight on the height of the D-layer, the propagation time (and hence the phase) of the received radio waves depends on whether any part of the radio path is in darkness. With Omega, because of the large distances likely to be involved, it may be dark at the receiving point and daytime at the transmitter, or vice versa. Alternatively, it may be dark at both transmitter and receiver and daytime half-way, or vice versa.

The propagation time and phase of the received waves therefore depend on the local time at the position of the ship and on the season. This last point will be clear: in arctic regions it is constantly dark in midwinter and light in midsummer.

Omega sky-wave correction tables

The LOPs printed on Omega charts are the result of a computation based on nominal conditions of propagation and earth conductivity. Because propagation conditions change continually, phase corrections (in centi-cycles) have to be applied for each station separately.

The US Naval Oceanographic Office publishes the *Omega Sky-wave Correction Tables*, which are sold by all authorised sales agents. *Figure 7.6* shows an extract from the tables for station A (Norway).

OMEGA SKYWAVE CORRECTIONS FOR 10.2KHZ   LOCATION 32.0 N   16.0 W   STATION A   NORWAY

DATE	GMT																								
	00	01	02	03	04	05	06	07	08	09	10	11	12	13	14	15	16	17	18	19	20	21	22	23	24
1-15 JAN	-49	-50	-50	-50	-50	-50	-50	-50	-26	-9	-15	-14	-12	-12	-13	-17	-23	-31	-40	-47	-49	-50	-50	-50	-49
16-31 JAN	-49	-50	-50	-50	-50	-50	-50	-50	-18	-10	-16	-13	-11	-11	-15	-20	-28	-38	-46	-49	-50	-50	-49		
1-14 FEB	-49	-50	-50	-50	-50	-50	-50	-47	-9	-16	-16	-12	-10	-9	-10	-12	-18	-26	-35	-45	-48	-49	-50	-50	-49
15-28 FEB	-49	-50	-50	-50	-50	-50	-50	-50	-21	-6	-16	-13	-9	-7	-7	-8	-10	-13	-21	-21	-43	-47	-41	-58	-58
1-15 MAR	-49	-50	-50	-50	-50	-50	-50	-44	-16	-13	-15	-10	-7	-5	-5	-6	-8	-11	-16	-27	-39	-46	-48	-49	-50
16-31 MAR	-49	-50	-50	-50	-50	-46	-31	-9	-15	-11	-7	-4	-4	-3	-4	-6	-8	-13	-21	-34	-45	-48	-49	-50	-49
1-15 APR	-49	-50	-50	-47	-37	-20	-12	-13	-8	-5	-3	-2	-2	-3	-4	-6	-10	-17	-27	-43	-47	-49	-50	-49	
16-30 APR	-49	-50	-50	-48	-42	-30	-13	-14	-11	-6	-3	-2	-1	-1	-2	-3	-5	-8	-14	-22	-39	-45	-48	-49	
1-15 MAY	-49	-50	-49	-43	-36	-24	-12	-13	-9			-1	-1	-1			-7	-12							
16-31 MAY	-48	-46	-47																						
1-15 JUN	-47																								

Figure 7.6 Omega sky-wave correction table. For example, on 10 January at 1100 GMT at grid point 32°N 16°W, the correction to the receiver readout for 10.2 kHz signals from station A (Norway) is −14 cec. In general, four such corrections are required for a fix

The correction parameters are the location of both station and ship, the date, and the time of day. There is therefore a separate series of tables for each station, consisting of a table for each 'grid point' (see below) in the station's coverage; in each table the whole hours of the day (GMT) are set out horizontally, and half-monthly periods vertically.

The grid system is as follows: a 4-degree grid for latitudes between 0° and 45°, a 6-degree grid between 45° and 60°, and an 8-degree grid between 60° and 90°. Corrections have been computed for each point of intersection of the grid lines, called a 'grid point'. The grid point to which each table applies is given at the top right (see *Figure 7.6*). The correction for the nearest grid point can be used, since the corrections only alter slightly for successive grid points. If, however, the ship is about half-way an interpolation can be made.

The phase correction for each station of the pair has to be obtained from the appropriate tables. The two figures are then combined into a single readout correction as follows. Suppose that the station-pair is A (Norway) D (North Dakota). It is 0400 GMT on 21 January. The position by dead reckoning is 31.30 N 16.22 W. According to the table in *Figure 7.6* the correction for A is −50 cec. From another table we find that the correction for D is −58 cec. The total correction for A − D is then (−50) − (−58) = +8 cec. In this case we should therefore add 8 cec to the readout. Note that to make the correct subtraction we should read A − D as A minus D, having placed the station letters in alphabetical order.

The sky-wave corrections given in the tables are normally negative, but can in exceptional cases be positive. By day the propagation tends to be stable, and corrections small, although conditions do vary slowly. At night the conditions tend to be constant but, the transit time having increased, corrections are higher than during the day. The transition periods from day to night and vice versa are of sufficient complexity that minor regular departures from the predicted corrections may be observed, particularly near the end of sunrise.

The corrections are, of course, determined by the conditions over the whole radio path, not just those at the ship's position. Note that if the ship is north or south of the transmitting station the radio path is entirely sunlit or entirely dark, and the corrections change very rapidly over a short period of time; if, on the contrary, the ship is east or west of the transmitting station the corrections alter more gradually.

From the sky-wave correction tables the general propagation conditions of a station can be discerned. According to *Figure 7.6*, for example, for the period from 1 to 15 January night conditions prevail between about 2000 and 0700 GMT, and day conditions between about 1000 and 1400 GMT. During the transitions the corrections alter rapidly.

Figure 7.7 is a graph of the corrections (for 16–31 January) during a night-day transition. With linear interpolation the correction at, for instance, 0730 GMT would be −33 cec. (Theoretically *Figure 7.7* should be a curve, but in practice straight lines and linear interpolation are adequate.) The use of Omega during the night-day and day-night transitions is not recommended, however. This is because the rapid changes during the transition may take place somewhat earlier or later, and *Figure 7.7* makes clear that this has great influence on the magnitude of the correction. Errors in the fix of several nautical miles can then be expected. *Tables therefore not only give the corrections but are also useful in discerning when the corrections of a station are not sufficiently reliable.*

As mentioned already, the waveguide effect may result in an extra

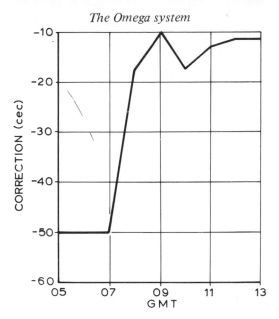

Figure 7.7 Graph of corrections (taken from Figure 7.6) for 16–31 January during the night-day transition. Because of their rapid changes, the corrections are not reliable during transitions

error at distances of less than about 650 n. miles from a transmitter. Although this error will not exceed two nautical miles, it is recommended to use another Omega station for that area.

Sometimes the printed corrections for a certain station's LOP in a given area are all too high or too low by the same amount. In such case a Notice to Mariners or other publication or message may read as follows: '32°N 16°W Norway subtract 10 cec from night readings'. Suppose this notice is received on 11 January, when the night conditions last (see *Figure 7.6*) from about 2000 to 0700 GMT. The correction of −50 cec during the night should be amended to −50 − 10 = −60 cec. By day the corrections are not subject to variations.

For the day-night and night-day transitions the following adjustments to the published corrections should be applied:

Time (GMT)	14	15	16	17	18	19	20	07	08	09
Computed sky-wave correction	−13	−17	−23	−31	−40	−47	−49	−50	−26	−9
Adjustment		0	−1	−3	−5	−7	−9	−10	−10	−5	0
Corrected value		−13	−18	−26	−36	−47	−56	−59	−60	−31	−9

208

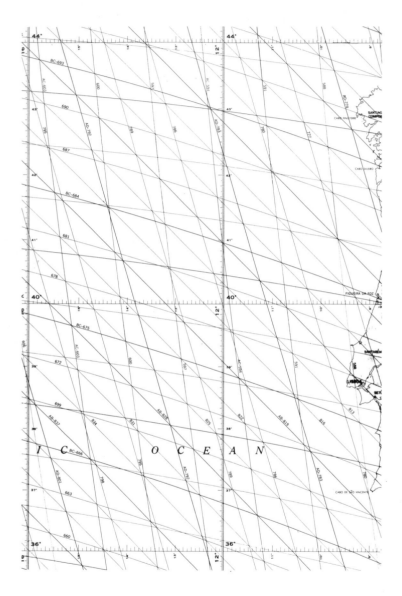

Figure 7.8 Omega chart (reproduced by courtesy of the US Naval Oceanographic Office)

Omega charts

Omega charts are published by the US Naval Oceanographic Office; a section of one of them is shown in *Figure 7.8*. The charts show only one in three of the Omega lines; each of those shown is marked with a lane number (the centreline is always 900), and one in three of those shown is also marked with the letters of the station-pair. The lines applying to the various station-pairs are printed in different colours, and no chart shows the lines for more than four station-pairs.

Since the Omega system is used not only by ships but also by aircraft there are land charts as well as sea charts. Note that Omega charts are only position-plotting charts, based on the Mercator projection, so each fix has to be transferred from the Omega chart to the normal navigation chart.

Omega tables

Instead of using special Omega charts the navigator can plot the Omega lines for himself on his navigation chart. For this purpose 'Omega Tables' are available that give the intersection points of Omega lines with lines of whole degrees of longitude and latitude.

This plotting procedure is rather time-consuming and the use of Omega charts is to be preferred. Moreover, plotting the lines from tables does not give them their correct curvature in the vicinity of a station. (Similar tables are available for the Loran system, but again experience has shown that the special Loran charts are preferable.) Therefore we shall not deal with Omega Tables in detail.

Accuracy

According to American publications the accuracy of a position fix by the Omega system is about 1 n. mile by day and about 2 n. miles at night. This is acceptable for ocean navigation but not for coastal navigation. For the latter there are other methods available all over the world, e.g. the Decca system or radio beacons. Possibly in the future accuracy will be improved when the magnitudes of corrections are better known.

Although the *absolute* accuracy of an Omega fix is only 1–2 n. miles, the *relative* accuracy (see page 45) is much better.

Differential Omega

It is possible to use the good relative accuracy obtainable with Omega to increase the absolute accuracy. For this purpose 'monitor stations' can be erected, for instance close to busy shipping areas (see *Figure 7.9*). Such a station determines its 'Omega position' at regular intervals, using each combination of Omega stations appropriate to the area. By comparing the results with its known true position, the monitor station is able to communicate exact and up-to-date corrections, for each station-pair, to ships

Figure 7.9 Differential Omega: the monitor station determines the difference between its 'Omega position' and its true position, and communicates this difference to ships and aircraft within a radius of 200 n. miles

and aircraft within a radius of 200 n. miles. This method of improving accuracy is called *Differential Omega*.

Identification of stations

In the proposed system of eight stations, the first station, A, begins its transmission (at a given frequency) a very short time after the last station, H, has finished its transmission; there is no distinctive break in the incoming signals. It is clear that this makes identification of stations difficult. Since the sequence of the eight stations is known, however, it is sufficient for one of the eight stations to be identified by the navigator. To do this he assumes that the nearest station is heard in his loudspeaker or headset more strongly than distant ones.

Each Omega receiver has an internal 'timer' or commutator, which generates a pattern that must be synchronized with the transmission format (*Figure 7.5*) of the Omega stations. For the purpose of synchronization, some receivers have a series of eight lamps on the front panel, which flash in sequence timed by the locally generated commutator pattern. The operator listens to the tones of the 10.2 kHz signals received from all the Omega stations, and aligns the light sequence with the signal sequence by releasing the commutator synchronization button at the exact moment that he hears the tone of a station he has identified and selected. Other receivers have only one lamp, which flashes as each signal is received. Sometimes use is made of a cathode-ray tube to compare the strengths of the stations on the screen. A more complete description of identification and synchronization is given later in the chapter.

Once the timer has been synchronized it will remain in that condition, and the required Omega stations can be selected and reselected as necessary. In order to avoid synchronization as much as possible, the receiver should be left on when the ship is in port or at any other fixed location.

A proposed method of automatic synchronization is as follows. Every Omega transmission cycle of 10 seconds starts at a point of time (clock time) that is a multiple of 10 seconds. As mentioned on page 219, an Omega receiver can be equipped with a clock that has sufficient accuracy to synchronize the timer with the correct Omega stations.

Radio waves from a station at a distance of, for instance, 9000 km will be received 9000/300 000 = 0.03 second after transmission. Because the Omega stations are at different distances from the receiver, the time between signals received from different stations can be shorter or longer than the 0.2 second gap between the signals as transmitted. However, the gap is long enough to prevent two stations' signals from being received simultaneously.

Identification with less than eight stations is easier than with the full eight stations because of the interruption in the cycle. In order to retain this advantage, the ranges of the eight stations will be such that no more than six stations can be received at any location in the world. There will then remain an interruption in the reception, to make identification easier.

A range as long as half the circumference of the Earth could cause difficulties for another reason. A ship at the far side of the Earth from an Omega transmitter would receive its signals via *two* paths of approximately equal length. However, the paths would not be of *exactly* the same length, and the phase of the resultant wave would be different from that of the wave following the shorter path (on which the charts are based).

Erratic identification

The nearest station does not under all circumstances give the strongest signal. This may give rise to a false identification. Take the case, for instance, of a ship in the Indian Ocean, sailing from South Africa to South Australia. The navigator wants to determine his position using the stations Norway (A), Réunion (E), South Australia (G) and Japan (H). (One of these stations is not yet operative). Stations E and G are nearer than the other two, and the distance to E is 350 n. miles less than to G.

The separation in longitude between E and G is very great, so propagation conditions (and therefore the 'diurnal shift') along the path from E to the ship will, at any given moment, be different from conditions along the path from G to the ship. A possible consequence of this is that between about 1600 and 2000 local time the nearest station E (Réunion) will come in weaker than G (South Australia). If unaware of this, one would be inclined, when listening to the stations, to take the strongest signal (which comes, in fact, from G) as being that of the nearest station, E. This would result in a false synchronization of the timer.

In *Figure 7.10* the schedule of 10.2 kHz signals is shown below the horizontal (time) axis as a series of *segments*, A to H. The signals received are indicated above the time axis, the height of each block being the field strength (corresponding to strength of sound).

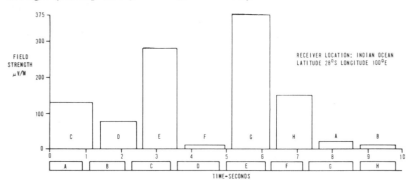

Figure 7.10 False synchronization: the internal timer is not correctly synchronized to the transmission format

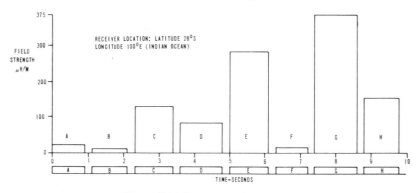

Figure 7.11 Correct synchronization

If the operator confuses stations E and G, as described above, he will initiate an incorrect timing or commutation in the receiver: the signals from station G will be treated as segment E (and similarly for the other stations). See *Figure 7.10*. The receiver will then make the wrong phase comparisons and give erroneous results. This can be prevented if the operator realises that: (*a*) in the case in question two strong signals can be expected; (*b*) the stronger of the two need not be that of the nearer station; (*c*) the *first* of the two closely-spaced strong signals is from E, because the interval between E and G is one second, and that between G and E is six seconds. *Figure 7.11* shows the correct synchronization.

A difficulty arises in this procedure if two equally strong signals are evenly spaced in the cycle (four seconds apart). Proper synchronization will be almost impossible, even for a skilled operator, because there will be no difference between the two intervals mentioned in (*c*) above.

Another problem is that a high level of atmospheric noise, precipitation static or local atmospheric disturbance causes the tones of an aural presentation to become obscured and a visual presentation to become difficult to read. Manual synchronization is therefore a weak point of the Omega system, especially in some areas, and there is a need for automatic synchronization.

The examples cited above demonstrate the problems of synchronization at sea; in some ports similar conditions occur, but the operator has more time to check the synchronization.

If the Omega receiver loses power for more than a few minutes the lane count can be lost or become confused, and resynchronization will be required. Therefore two separate power sources should be available, one of them a battery that can automatically take over if the normal supply fails. In some receivers a buzzer gives an alarm signal when this happens. An Omega receiver's consumption of power is very low, about 40 to 100 watts.

Sperry Omega receiver SR-500

Figure 7.12 shows the Sperry Omega receiver type SR-500 with the following controls:

1. *Volume control.* Controls level of audio output to speaker and/or phones. When rotated fully counterclockwise ('Alarm Disable'), resets and disables alarm. (Alarm will sound when power to the receiver is too low.)
2. *Synchronization pushbutton.* Starts synchronization cycle with the segment indicated by the left-hand LOP 1 switch.
3. *Segment indicator lamp.* Indicates internal timer sequence, with brighter illumination for segment selected by left-hand LOP 1 switch.
4. *Ready indicator lamp.* Lights to show that warm-up period has expired and Omega receiver is ready for operation.
5 and 6. *LOP 1 left and right channel switches.* Select first and second segments, respectively, of LOP 1.
7 and 8. *LOP 2 left and right channel switches.* Select first and second segments, respectively, of LOP 2.
9. *LOP 1 lanes indicator.* Indicates lane and centilane of LOP 1. Starting from the left, the first four wheels, each single-digit, indicate the lane number and can be adjusted manually by means of the knob above the counter. The fifth wheel, which is double-digit (00 to 99), indicates the centilane number: *never try to adjust this wheel*, which is

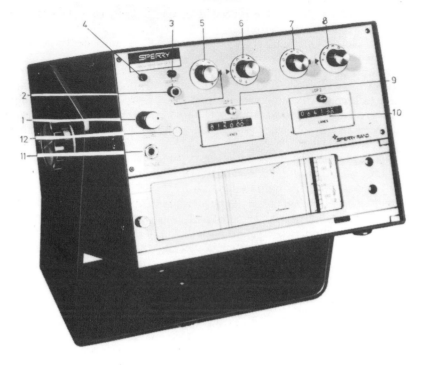

Figure 7.12 Sperry Omega receiver SR-500

driven by the receiver according to the Omega signals received, and is mechanically coupled to the lane wheels.

10. *LOP 2 lanes indicator.* Indicates lane and centilane of LOP 2.
11. *Phones connector.* Permits monitoring of Omega receiver with a set of earphones.
12. *Speaker.* Permits monitoring of Omega segment tones.

Synchronization of Sperry receiver

To synchronize the timer, follow these instructions:

(*a*) Adjust the volume control. Unless static is severe in any location, at least three stations should be heard via the speaker or headset.

(*b*) Identify each tone with the corresponding Omega station by using available knowledge of the station locations, the operating area, and the standard Omega transmission format (*Figure 7.5*).

(*c*) When you are sure you can identify each tone with a specific station, set the LOP 1 left channel switch (*Figure 7.12* item 5) to the segment that corresponds to the station tone that is the easiest to hear and recognise.

(d) Notice that the 'Segment' lamp (*Figure 7.12* item 3) flashes on each timer segment, and that it flashes brighter on one segment than on the others. This brighter segment corresponds to the one you have selected with the LOP 1 left channel switch.

(e) Depress the 'Sync.' pushbutton (*Figure 7.12* item 2) and listen for the tone of the station you have selected in step (c). *Release the 'Sync.' pushbutton at the exact time that you hear your station* (or as close as possible). If you have synchronized the receiver properly, the 'Segment' lamp will now flash brightest at the same time that you hear the tone of the station you selected in step (c).

The absence of the 10.2 kHz Omega signal in one or more segments selected by the LOP switches will cause the alarm to sound. Such an alarm indicates that a selected Omega signal has been lost or the Omega receiver is not synchronized properly. (The alarm also sounds when the precision crystal oscillator in the receiver requires alignment.)

Recorder of Sperry receiver

The recorder (lower part of front panel in *Figure 7.12*) displays a centilane-versus-time record of each LOP selected on the receiver. The chart speed is half an inch per hour. Time marks, which clamp the chart trace to 0 cel for two minutes, are provided every 30 minutes to ensure an accurate time reference for the chart record.

Determining lane count while in port

1. Check that the receiver is properly operating and synchronized.
2. Using the Omega chart, determine the two LOPs for the ship's position and select the two segments for LOP 1, for instance A and D, using the two LOP 1 channel switches.
3. Look up the sky-wave corrections for both A and D and determine their difference. For instance, if the correction for A is −30 cec and that for D is −52 cec, the difference is (−30) − (−52) = −30 + 52 = +22 cec.
4. *Subtract* this difference of +22 cec from the LOP 1 chart reading. The result is what the receiver ought to indicate on the LOP 1 lane counter.
5. If necessary, adjust the lane counter to the expected *lane* number. Never attempt to adjust the *centilane* number; the only thing to check in this respect is that the expected centilane reading is within 25 cel divisions of the actual value indicated on the lane counter. For example, if the expected LOP reading is 783.95 and, disregarding the

lane count, the actual centilane reading is 15, then the lane number should be set to 784.15 and not 783.15, because the difference from the expected reading should be 25 cel or less.

6. Follow the same procedure for LOP 2.

Making an Omega fix

For making an Omega fix at sea Sperry issues special data sheets, an example of which is shown in *Figure 7.13*. The procedure is as follows:

(*a*) Enter on line (1) of the sheet the local time and date, and on line (2) the GMT.

OMEGA DATA SHEET

(1) TIME _____ LOCAL   DATE _____

(2) TIME _____ GMT

(3) LATITUDE _____

    LONGITUDE _____

(4) LOP 1 SEGMENTS	_____		LOP 2 SEGMENTS	_____
(5) SWC 1 (+)	X X X . ___		SWC 1 (+)	X X X . ___
(6) SWC 2 (−)	X X X . ___		SWC 2 (−)	X X X . ___
(7) TOTAL CORRECTION	X X X . ___		TOTAL CORRECTION	X X X . ___
(8) LOP 1 READING	_____ . ___		LOP 2 READING	_____ . ___
(9) LOP 1 CORRECTED	_____ . ___		LOP 2 CORRECTED	_____ . ___
(10) OMEGA LATITUDE	_____			
OMEGA LONGITUDE	_____			

NOTES

Figure 7.13 Data sheet for Sperry Omega receiver

(*b*) Enter on line (3) the appropriate latitude and longitude values from a previous fix, if available, or make an appropriate fix by plotting the lane-counter readings on an Omega chart and use the resulting latitude and longitude values.

(*c*) On line (4), enter the LOP 1 and LOP 2 segment pairs.

(*d*) Look up the corrections for the first segments of LOP 1 and LOP 2 in the *Omega Sky-wave Correction Tables*. Enter the values on line (5).

(*e*) Look up the sky-wave corrections for the second segments of LOP 1 and LOP 2, and enter them on line (6). When making these entries, change the sign of each number.

(*f*) Now add lines (5) and (6) algebraically (for each LOP) and write the sum on line (7).

(g) Enter the LOP 1 and LOP 2 readings from the lane counters on line (8).

(h) Finally, add lines (7) and (8) algebraically and enter the result on line (9). Remember to take the sign on line (7) into account.

(i) Plot the corrected LOPs on an Omega chart. Determine the latitude and longitude values from the plot and enter them on lines (10). This represents your present position.

Direct measurement of distance from transmitter

Suppose that a ship has on board an atomic oscillator similar to the one in the Omega transmitter, and that the signals generated by the two oscillators are exactly matched in frequency and phase. If one imagines the ship and transmitter to be at the same position, each wave transmitted arrives on board at the moment of transmission, and the phase difference between the received waves and the internally generated waves is zero.

If the ship moves away from the transmitter, the radio waves arrive later and a phase difference is measured. At a distance of one wavelength (30 km with Omega) this difference is 360° or 100 cec. The locus of the position of the ship is then a circle with a radius of 30 km and with the transmitting aerial as its centre.

If the ship keeps on moving away the phase difference increases again from zero. When it again reaches 360° or 100 cec, the distance from the transmitter is 2 × 30 km = 60 km. This gives a new circular LOP, concentric with the first. The lane (or, since it is annular, 'ring') between these two LOPs is subdivided by intermediate LOPs corresponding to each phase difference between 0° and 360°. Charts can be overprinted with these LOPs. If the receiver incorporates a pointer that makes one revolution per 360° change in phase difference, and also a device to count the number of pointer revolutions, the distance from the transmitter can be determined.

Of course, it is not necessary for the ship's and the transmitter's positions to be the same at the outset. If the navigator knows the ship's position by other means (e.g. when lying in port), he can read from the chart the number of the appropriate ring and the phase difference (0–360°). He can now manually set the indicator to the correct reading; each change in the reading will then correspond to an increase or decrease in distance from the transmitter.

Note that an LOP is obtained using only one transmitter. With two transmitters, and a duplication of the equipment on board, a radio position fix is obtained.

It may be that in the future some such system will be introduced. Special transmissions for this and/or other purposes could take place during the time that the transmitter is not operating on one of the three

Omega frequencies. Station B, for instance, could transmit for about five seconds at frequency f_2 (see *Figure 7.5*) after its $11\frac{1}{3}$ kHz transmission.

Advantages and disadvantages of Omega

Comparing Omega with other hyperbolic systems, the following points are to the advantage of Omega:

(a) It can be used all over the world on land and sea, and below the sea, so changing during the voyage or flight from one system to another is not necessary. Of course one has to change from one pair of stations to another, but there is no break in the position fixing.

(b) All the stations together make use of only three frequencies, in a frequency band that is used by only a small number of transmitting stations.

(c) In the future, at every location in the world the number of receivable station-pairs will be sufficient to provide LOPs with a good inter-section angle.

(d) Accuracy will probably be improved in the future as a result of better predictions of sky-wave corrections.

(e) The relative accuracy, important for some applications, is reasonably good.

(f) The user does not contribute to the cost of the system.

On the other hand there are the following disadvantages:

(a) The *absolute* accuracy is, at present, not so good.

(b) There is the possibility of lane slip.

(c) In general four stations are required for a position fix, so one has to look up four sky-wave corrections and apply them.

(d) It is difficult to identify the stations, especially in certain areas.

(e) At distances shorter than 650 miles from a transmitter the accuracy can be less than at longer distances.

Displays

The receivers display the lanes and centilanes in one or more of the following ways:

(a) on a digital trip-counter, as in the Sperry receiver described;

(b) on a meter;

(c) by recording on a strip of paper (recorder); the paper strip (printed with a scale) moves at a constant velocity, while the pen is displaced perpendicularly to the paper movement by the received signals;

(d) by printing the lanes and centilanes on a strip of paper.

The disadvantage of method (c) is that the thickness of the line can make the readings less accurate. Method (a) seems to be the best for a rapid reading, preferably coupled with recording of the centilanes in order to be able to check on past readings. Some receivers display two readings simultaneously, others only one at a time.

Auxiliary apparatus

There are various pieces of auxiliary equipment designed for use with an Omega receiver:

(a) a clock that can be used as a very accurate chronometer (accuracy 0.1 second);

(b) an apparatus that automatically converts the lanes and centilanes to longitude and latitude;

(c) an apparatus that automatically charts the position, based on LOP readings;

(d) with Omega receivers forming part of an integrated navigation system (Chapter 10), a computer that calculates the sky-wave corrections as a function of (among other things) the height of the reflecting layer.

8

The navigation satellite system

Principle

Position fixing by means of the present satellite system is based on measurement of the frequency alterations that arise if the receiver on, for instance, a ship is tuned to the transmitter of an *artificial earth satellite* moving relative to the receiver (the Doppler effect, described in Chapter 1).

Figure 8.1(a) shows the paths of four transmitters A, B, C and D. The transmitters, on different paths but moving at an equal and constant speed, move relative to a receiver at O. Transmitters B and C are at the shortest distance from O simultaneously (time t_1), transmitter D somewhat later (time t_2). In *Figure 8.1(b)* the frequency received at O is plotted as a function of time, to show the Doppler effect.

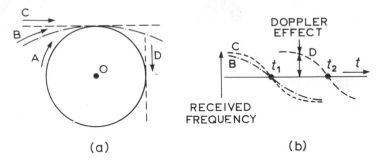

(a) (b)

Figure 8.1 At the receiver O, there is no Doppler effect from transmitter A; the frequency received from transmitter B changes more slowly than that from C; the curve for the frequency received from D is equal to the curve for C

For transmitter A, with its circular orbit around O, the distance between transmitter and receiver does not change and there is no Doppler effect. The frequency received is therefore equal to the transmitter frequency and is indicated by the time-axis. For transmitter B, the distance from O changes more slowly than for transmitter C, resulting in C's steeper

Doppler curve at t_1. (Times t_1 and t_2, when the transmitter-receiver distance is smallest and the received frequency equals the transmitter frequency, are *times of passing*.) The curve for transmitter D is equal to that for C, but the time of passing, t_2, is later than C's.

A satellite is a celestial body that accompanies a planet in its orbit and moves around it according to the laws of Keppler. There are numerous artificial Earth satellites, put into orbit for very different purposes: there are satellites for communication, for weather monitoring and for position-fixing purposes. In 1977 special satellites for ships' communications became operational.

On a satellite there are two principal forces, acting in opposite directions: the *centrifugal* force (proportional to the square of the velocity) and the *gravitational* force (inversely proportional to the square of the distance to the centre of gravity of the Earth). For each velocity of the satellite that is sufficiently great (but not too great), the satellite follows an elliptical orbit determined by the balancing of the centrifugal and gravitational forces. Since orbit height and speed determine the time of one revolution, there is also a direct relation between this time and the height.

The first artificial Earth satellite, Sputnik I, was launched by the Russians in 1957. The Americans Guier and Weiffenbach of the Johns Hopkins Applied Physics Laboratory received the radio signals from this satellite; as one would expect, owing to the Doppler effect the received frequency gradually decreased as the satellite approached and receded. Further investigation of the frequency alterations enabled them to determine the orbit of the satellite. This very important discovery (astonishing considering that the Americans made it purely by observing a Russian satellite) was based on existing knowledge and technology, viz.

(a) the laws of Keppler and Newton regarding the orbits of satellites;
(b) the possibility of recording Doppler frequency shifts very accurately
   as a function of time.

Orbit of navigation satellites

We will now consider the American *Navy Navigation Satellite System* (NNSS), better known under its former name *Transit*. Since 1964 these satellites have served for position-fixing purposes by means of the Doppler effect. In 1967 the system was released by Vice-President Humphrey for non-military purposes. On that occasion he said: 'This system enabled fleet units to pinpoint their positions anywhere on Earth. The same degree of navigational accuracy will now be available to our non-military ships.'

The orbit of the satellites is an ellipse with such a small eccentricity that it is almost a circle. Its height is about 1100 km; on lower orbits air drag would be too great, on higher ones the Doppler frequency shift would be too small. The velocity of the satellites is on average 7.3 km/s (about

26 000 km/h); they orbit the Earth in about 1¾ hour, thus making about 13½ orbits a day.

The orbit is almost polar, i.e. the plane of the orbit passes near the two Earth poles. Basically this plane has a fixed position within a coordinate system based on the fixed stars, and therefore moves parallel to itself (due to the Earth's orbital movement around the sun) but does not rotate in that system. For certain reasons there is, however, a slight *precession*, i.e. a small rotation of the orbit-plane.

Owing to the Earth's rotation, each point on Earth passes beneath any of these orbits at a velocity proportional to the cosine of its latitude. The projection of a satellite's path on to the Earth's surface is not therefore directly north-south but has a *westerly component*, the size of which is dependent on the cosine of the latitude. The same satellite passes over a given place about once every 12 hours.

Figure 8.2 Six polar satellite orbits

The satellite's signals can be received only if it is above the horizon. At places where its maximum elevation angle is small, the satellite appears only just above the eastern or western horizon.

Figure 8.2 shows the approximate orbits of six navigation satellites existing in 1969. In February 1973 there were five satellites plus a 'rogue' one. *Figure 8.3* shows the orbit configuration of the five satellites on 5 February 1973, in a view from above the Earth with the North Pole as the centre. The rogue satellite is no longer operative but cannot be permanently turned off; as long as its signals do not interfere with a useful satellite, however, this does not cause problems. It is reported that satellite 30180 is no longer used, because its signal strength has for some time been too low. In 1975 satellite 30200 was launched.

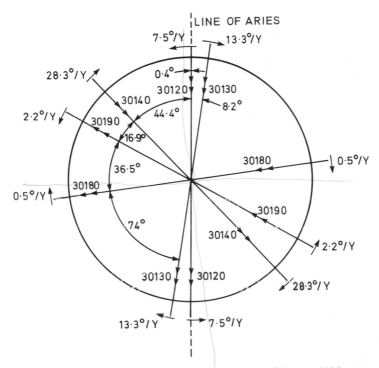

Figure 8.3 Orbital separation of the five satellites on 5 February 1973; centre of circle is the North Pole. Precession of each orbit is indicated in degrees per year. 'Rogue' satellite is not included in diagram

At or close to the poles all satellites can be received and made use of; at the equator the number of satellite passes per day is smallest. With the present five satellites it can be said that the time between passes, averaged for the whole world, is 2.4 hours. The *maximum* time between useful passes was four hours at the beginning of 1971. It should be noted that, if two satellites pass over a receiver simultaneously, they can cause mutual signal interference.

Each satellite is equipped with electronic apparatus chiefly consisting of a radio transmitter (operating on two frequencies simultaneously), a receiver, an aerial-system and a digital memory. Its power is generated by means of solar cells (*Figure 8.4*) on the four 'wings' that are unfolded after

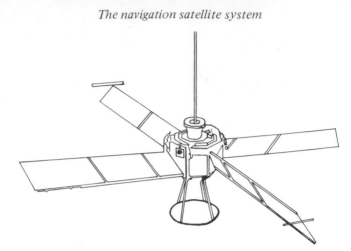

*Figure 8.4 NNSS satellite, with solar cells on the 'wings' for con-
verting sunlight into electrical energy. The long rod above
establishes the gravity gradient stabilisation so that the 'lampshade' aerial below is directed towards the Earth*

launching. As long as the satellite is not in the shadow of the Earth the
solar cells charge a battery of nickel-cadmium cells. (There are also satellites that obtain their power from a thermal radio-isotope source.) For
the purpose of directing the 'lampshade' aerial on the underside of the
satellite towards the Earth, there is a long rod, extending from the top of
the satellite, that automatically points away from the Earth.

Deviations in the elliptical orbit

In order to determine the position of a point on the Earth, it is necessary
to know the position of the satellite in its orbit during the Doppler-shift
measuring process. All deviations from the nominal orbit must therefore
be known. Apart from the very slow precession, there are four causes of
orbit deviation:

(*a*) irregularities of the Earth's gravity field, of an extent not exactly
known;

(*b*) the air drag experienced by the satellite (even though, at the height of
about 1100 km, the atmosphere is extremely rarefied);

(*c*) the pressure exerted on the satellite by solar radiation;

(*d*) the attraction of other celestial bodies (for instance the moon).

Of these causes the first creates the greatest difficulty. The field of gravity
shows irregularities because the mass of the Earth is not equally distri-
buted. Nevertheless it has been possible to determine these irregularities
to some extent in the following way.

If one knows exactly a satellite's deviation from its nominal orbit at any given time, one can determine one's position on Earth by making Doppler frequency-shift measurements; this is what the navigator needs to do for position-fixing. Conversely, therefore, one can calculate the satellite's deviations as a function of time by making Doppler measurements from a number of *known* positions on Earth. (The calculations are more complex, however.) By extrapolation, the deviations from subsequent orbits can also be determined.

American scientists succeeded in calculating a gravity model of the Earth by observing the Doppler shifts of NNSS satellites, and also by tracking other satellites (by means other than Doppler observations) from a global net of tracking stations at known positions. Terrestrial information about gravity anomalies was also used. As further research is carried out and more information becomes available, the gravity model is continually improved.

A way of showing the Earth's gravity field is by means of a *geoidal chart* (*Figure 8.5*). The geoid is the shape of the Earth determined by neglecting relief (mountains etc.); it is an imaginary surface that coincides with the mean sea level. Owing to gravity anomalies caused by unequal

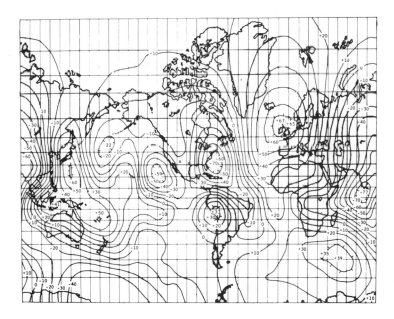

Figure 8.5 Geoidal chart showing deviations (in metres) of the Earth's shape (represented by mean sea level) from the shape of a reference ellipsoid. Thus the level in the North Sea is about 65 m higher, and that near Sri Lanka about 77 m lower, than the reference level. The geoid is defined by the gravity model used since 1973

mass distribution, the geoid deviates from the Earth ellipsoid. *Figure 8.5* shows these deviations, in metres, with respect to an Earth ellipsoid with half a long axis of 6378.166 km and an eccentricity of 1/298.3.

The 'geoidal height', as indicated in *Figure 8.5*, for the position of a ship or other point of observation must be supplied to the computer to which the satellite receiver is connected. For a point on land at unknown elevation, height as well as latitude and longitude would have to be calculated. Since at sea the height (approximately the geoidal height) is known, only latitude and longitude have to be found; this can be done more accurately with one fewer unknown. The computer needs to know the height difference between geoid and ellipsoid because its position calculations have to be based on the reference ellipsoid.

Storage of predicted orbit deviations in satellite memory

In the United States four fixed stations (tracking stations) track the satellites during their passes by means of Doppler measurements (*Figure 8.6*). The results of the measurements are transmitted to a computing centre. This centre also receives time signals from the naval observatory.

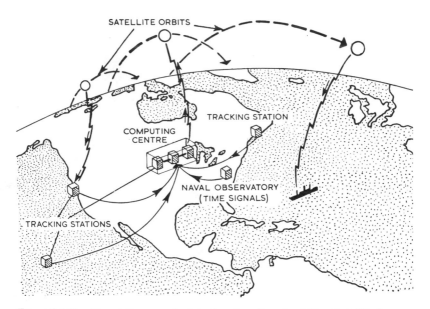

Figure 8.6 Tracking stations supply data, based on Doppler shifts of satellite frequencies, to the computing centre, which calculates orbit coordinates for the next 16 hours. The injection station transmits this information to the satellites, which store it and transmit it regularly to system users

The computers in the centre *calculate and predict the orbit coordinates for the following 16 hours.* An injection station transmits these coordinates to the satellite. They are stored in its memory and transmitted by the satellite to its users (for instance, ships). Every 12 hours there is a new injection of data. It is possible to indicate the current orbit of a satellite within an error of 10 to 15 metres and to predict the deviations of future orbits within an error of 40 metres.

The satellite signals must pass the ionosphere in order to reach the lower troposphere and the earth. As frequencies lower than roughly 40 MHz are reflected by the outer surface of the ionosphere, the satellite frequency must be higher than 40 MHz. In fact two frequencies are transmitted simultaneously: one of about 400 MHz (399.968 MHz exactly), the other of about 150 MHz (149.988 MHz exactly); both are multiples (harmonics) of a fundamental frequency of 49.996 kHz.

Doppler curves

Suppose a satellite starts transmission at A of its orbit (*Figure 8.7*) and stops at B. The velocity of the satellite is constant, and there is a stationary receiver at O. Because the satellite is approaching O, every wave generated

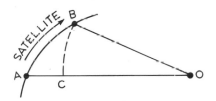

Figure 8.7 The number of cycles transmitted by the satellite between A and B is equal to the number received at O. Since the satellite is moving closer to the receiver, however, this number of cycles arrives at O within a shorter period; thus the received frequency is higher than the transmitted frequency

is shortened in length. Although the number of waves received must equal the number generated, they are received over a shorter period. The frequency is consequently higher than if the distance between the satellite and the receiver had not changed; this is the Doppler effect, described more fully in Chapter 1. As will be shown below, the difference in distance AO − BO can be calculated from the increase in the received frequency and from other known quantities.

Figure 8.8 shows some Doppler curves similar to those in *Figure 8.1*. Curve A applies to an observer A, situated at a certain distance west of the terrestrial projection of a south-going satellite orbit. (This projection is not exactly north-south because of the Earth's rotation, as mentioned previously. However, in view of the much shorter period of the satellite's orbit (1¾ hour) in relation to that of the Earth's rotation (24 hours), we shall assume for convenience's sake that the projection is north-south.) Curve B applies to an observer B, closer to the orbit projection. For observer C, at the same distance from the orbit projection as A but further to the south, the moment of passing is delayed but the Doppler curve is the same as that for A.

It appears from the Doppler curves that the slope or rate of change of the frequency, especially at the centre (the moment of passing), depends on the distance between observer and orbit projection. This distance is the

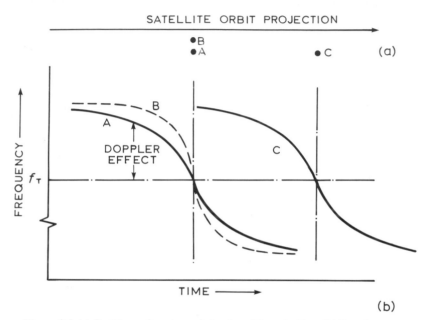

Figure 8.8 (a) Positions of receiver relative to orbit projection; (b) Doppler curves for those positions

difference in *longitude*. The steeper the curve at the moment of passing, the smaller the difference in longitude. Also, the moment at which the received frequency is equal to the transmitted frequency f_T defines the *latitude* of the observer. So there are two important features of a Doppler curve:

(a) the point at which it crosses the f_T line defines the observer's position along the orbit projection, i.e. latitude;

(b) its slope, especially at the centre, defines the difference in longitude between observer and orbit projection.

Derivation of Doppler-count formula

Accurate measurement of the *instantaneous* received frequency meets with difficulties of a technical nature, and so another method is used. In the following derivation only the 400 MHz satellite frequency is considered, not the 150 MHz frequency.

The received frequency f is mixed with a frequency f_g generated in the receiver; f_g is higher than the highest value that f can attain and amounts to 400 MHz. The number of *beats* per second (the beat frequency, see Chapter 1) originating from this mixing is equal to the difference $f_g - f$ and is therefore always positive. The beat frequency is considerably lower than the received frequency and is on average $400 - 399.968 = 0.032$ MHz $= 32$ kHz.

What we want to know is the number of beats during an interval of two minutes, the beginning and end of which are marked by time signals from the satellite transmitter at t_1 and t_2. The number of beats during a certain period (the *Doppler count*) can be calculated in the following way. (Readers not familiar with mathematical integration may omit this description.)

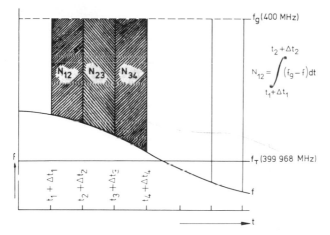

Figure 8.9 Areas N_{12}, N_{23}, etc. are proportional to the number of beats obtained in the receiver, during each two-minute transmission period, by mixing the received frequency f with the generated frequency f_g. From this number (the 'Doppler count') a locus of the ship's position can be obtained

In *Figure 8.9* a Doppler curve has been sketched. Each of the hatched areas has a variable height equal to the instantaneous beat frequency $f_g - f$ and a width equal to a time of two minutes. By multiplying the frequency in cycles per second by the time in seconds, the number of cycles N_{12}, N_{23} etc. is obtained. Each of the hatched areas represents, therefore, the number of beats during two minutes. However, on account of the variability of the beat frequency we have to integrate instead of multiply. For the present we assume that the observer does not move on the Earth's surface.

The time signals transmitted at t_1, t_2, etc. arrive somewhat later at the receiving aerial, i.e. at $t_1 + \Delta t_1$, $t_2 + \Delta t_2$, etc.

$$N_{12} = \int_{t_1 + \Delta t_1}^{t_2 + \Delta t_2} (f_g - f)\mathrm{d}t = \int_{t_1 + \Delta t_1}^{t_2 + \Delta t_2} f_g \mathrm{d}t - \int_{t_1 + \Delta t_1}^{t_2 + \Delta t_2} f \mathrm{d}t$$

Because f_g is assumed to be constant:

$$N_{12} = f_g \left[(t_2 + \Delta t_2) - (t_1 + \Delta t_1) \right] - \int_{t_1 + \Delta t_1}^{t_2 + \Delta t_2} f \mathrm{d}t$$

$$= f_g \left[(t_2 - t_1) + (\Delta t_2 - \Delta t_1) \right] - \int_{t_1 + \Delta t_1}^{t_2 + \Delta t_2} f \mathrm{d}t$$

The number of cycles leaving the transmitter (at frequency f_T) in two minutes must be equal to the number that arrive in due course at the receiver.

$$\int_{t_1}^{t_2} f_T \mathrm{d}t \equiv \int_{t_1 + \Delta t_1}^{t_2 + \Delta t_2} f \mathrm{d}t$$

If we assume that f_T is also constant, the number of cycles is $f_T(t_2 - t_1)$ so that:

$$N_{12} = (f_g - f_T)(t_2 - t_1) + f_g(\Delta t_2 - \Delta t_1)$$

Now $f_g = c/\lambda_g$, where c = the velocity of propagation and λ_g is the wavelength. As $c \times \Delta t$ is equal to the path covered by the radio waves in Δt seconds, we may assume that:

$$f_g (\Delta t_2 - \Delta t_1) = \frac{c}{\lambda_g}(\Delta t_2 - \Delta t_1) = \frac{1}{\lambda_g}(S_2 - S_1)$$

where S_1 and S_2 are the distances between the observer and the satellite at t_1 and t_2. Thus, finally:

$$N_{12} = (f_g - f_T)(t_2 - t_1) + \frac{1}{\lambda_g}(S_2 - S_1)$$

We can assume that f_g, f_T and λ_g are known; $t_2 - t_1 = 120$ seconds, and N_{12} is the Doppler count measured by electronic methods. Therefore with this formula it is possible to calculate the unknown difference in distance, $S_2 - S_1$. It can be concluded from the formula that an error of one beat in the count N_{12} at a frequency f_g of 400 MHz (λ_g = ¾ metre) corresponds to an error of ¾ metre in the difference in distance $S_2 - S_1$.

A satellite pass lasts long enough to make five to eight counts of two minutes possible.

Computer calculation of ship's position

Naturally, movement of the receiver over the Earth's surface (for instance aboard ship) changes its distance from the satellite and thus influences the Doppler count. The course and speed of the ship must therefore be supplied to the computer that performs the calculation. For the moment, however, we assume reception aboard a stationary ship.

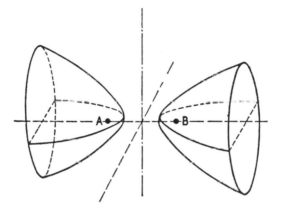

Figure 8.10 Hyperboloid: every point on its surface has the same difference in distance from the two fixed points A and B (the foci)

A *hyperboloid of revolution* (*Figure 8.10*) is obtained by rotation of a hyperbola about its axis. (In previous chapters we have considered the hyperbola as a *single* curve. Mathematically, however, it is a *symmetrical pair* of curves, and is so treated in what follows.) Every point on the surface of the hyperboloid has the same difference in distance from two fixed points, the *foci* A and B in the figure. If points A and B are known, a certain hyperboloid exists for a known difference in distance, and conversely.

We now take a coordinate system based on free space (fixed stars), rather than a terrestrial system. We take the locations of the satellite at times t_1 and t_2 as the hyperboloid foci A and B. The distance between A and B is the path covered by the satellite in two minutes, and amounts to about 850 km at the average velocity of 7.3 m/s.

If, on the basis of a Doppler count, we know $S_2 - S_1$, i.e. the difference in the distances between the observer and each of the two foci, the observer's position must be somewhere *on the surface of the hyperboloid for this difference in distance*. We know, moreover, that the observer on board a ship is on the surface of the Earth. (For convenience's sake we assume here that he is on the surface of a reference ellipsoid that can be taken as the approximate mathematical shape of the Earth.) Thus *a locus of the observer's position is the pair of lines of intersection of the hyperboloid and the Earth ellipsoid*.

A second locus can be obtained by means of a second Doppler count of two minutes, immediately following the first one and from the same satellite. Because the satellite is further on in its orbit there are two new foci, the new focus S_1 coinciding with the previous focus S_2. Another hyperboloid and another pair of lines of intersection with the Earth's surface are thus obtained.

The two loci will intersect at more than one point. The digital computer, connected to the receiver, selects the correct point of intersection as the position of the ship. For this purpose it is necessary, when setting up the receiver, for an initial estimate of the ship's latitude and longitude to be fed into the computer. The distances between the points of intersection are so large that this estimated position need only be accurate to within 3° of latitude and 3° of longitude.

Although the loci of the ship's position lie on hyperboloidal surfaces, the system described is not considered to be a hyperbolic system of position fixing like, for instance, the Loran system; among other things, the foci of the hyperboloid are in different positions for different satellites and for different orbits of the same satellite, so the intersection lines of hyperboloid and Earth cannot be indicated on charts.

Up to now we have taken the difference in frequency $f_g - f_T$ as being known. Both frequencies, generated in receiver and satellite transmitter respectively, indeed have a very great stability. That in the satellite transmitter is 1 in 10^{11}, which is exceptionally stable considering that the satellite is exposed to very strong solar radiation that has not been absorbed by the atmosphere. In addition, there is no cooling by air conduction but only by metal conduction and by radiation. The satellite temperature in sunlight is therefore very high, but in the shadow of the earth it sharply decreases.

In spite of their stability, the frequencies f_T and f_g of satellite and ship's receiver will gradually drift apart; therefore the computer considers the difference in frequency $f_g - f_T$ as an unknown quantity. So a third Doppler count is necessary, to make calculation of the position possible. There are then three unknowns and three independent equations.

Doppler counts are made for as long as the satellite can be received (say, five or six two-minute counts). The computer then calculates the range differences between the two known satellite positions (the foci of the hyperboloid) and the estimated ship's position, and compares these differences with those based on Doppler counts. The ship's actual position is then obtained by finding those values of latitude, longitude and $f_g - f_T$ that give the best agreement between the calculated range difference and the Doppler-measured range difference.

Let us now assume that the ship is moving. The speed and course must be provided to the computer so that it can take into account the ship's movement during each two-minute count. In other words, the computer finds the position on Earth (based on the ship's movement during the measuring interval) that best fits the range differences measured by Doppler counts. This computer calculation is repeated a few times, starting each time with the longitude and latitude found by the previous calculation. The differences between calculated and observed Doppler counts (residuals) thus steadily decrease. i.e. the solution *converges*. Usually no more than three or four iterations are necessary. A small computer will need one minute at most for the calculations.

Data included in satellite signals

The signals that are used for making Doppler counts are simultaneously used for transmitting navigational and other information. The latter function does not affect the former because phase modulation is applied. The data is transmitted simultaneously on 150 and 400 MHz.

A satellite message of two minutes duration consists of 26 lines of six words each. Only word 6 of each line contains navigation information; the rest is classified 'secret' and is not decoded in navigation receivers. Each word consists of 39 bits, of which only 32 (or 36) are used.

ACCURACY

There are five factors influencing the accuracy of the fix:
(*a*) an error in the ship's course and/or speed provided to the computer;
(*b*) anomalies of signal propagation;
(*c*) an error in the antenna height provided to the computer;
(*d*) inaccuracy of apparatus on board ship;
(*e*) errors in the predicted orbit deviations.

Influence of course and speed

The distance covered by a ship travelling at 18 knots during a pass of 10 minutes (five counts of 2 minutes) is 3 n. miles or 5.5 km. This is a very large distance compared with the accuracy of the system, and must not be neglected. The ship's course and speed can be supplied to the computer either manually or automatically.

Figure 8.11(a) shows the errors in longitude and latitude as a function of the satellite's elevation at closest approach (at a latitude of about 30° north) caused by an error of one knot in the *northern* direction in the speed supplied to the computer. This error is the northern component of the speed error vector, which is independent of the ship's course. Take a very simple example: the course supplied and the true course are both south, the supplied speed is 15.5 knots and the true speed is 16 knots. The *north* error is then $16 - 15.5 = +0.5$ knot. Normally, of course, the speed error vector consists of a northern and an eastern component.

The longitude error in *Figure 8.11(a)* is greater than the latitude error. This can be explained as follows. If the ship and satellite are both moving north or south, the Doppler count is less than if they are moving in opposite directions. So the consequence of an erroneously provided speed in the north-south direction is that a ship will receive more or fewer radio waves than the computer will calculate on the basis of the erroneous speed. The increase or decrease in the number of radio waves will be greater for speeds in the north-south direction than for speeds in the east-west direction. The computed Doppler curve based on the erroneous speed will then be either too steep or not steep enough, as the case may be. This will result mainly in a longitude error, as can be seen from *Figure 8.8*.

Figure 8.11(b) shows the errors for each error of one knot in the *eastern* direction in the speed supplied to the computer (again for a latitude of about 30° north). *In principle* the latitude error is now greater than the longitude error, for the following reason. An important contribution to the Doppler shift is made by the Earth's rotation, which gives each point on Earth a speed proportional to the cosine of its latitude. The computer takes the rotation into account in its calculations. An error in the eastern component of the ship's speed will give the impression that the Earth's rotation is faster or slower than the speed corresponding to the latitude of that point. This will result in a satellite fix giving a latitude nearer to or further from the equator than in reality. In other words, the east-west speed error should influence the latitude more than the longitude.

A careful study shows, however, that this statement is only partly true. The satellite's path is not exactly north-south but has, as a result of the Earth's rotation, a westerly component in its movement relative to the observer. The steepness of the Doppler curve is therefore also slightly affected by errors in the east-west speed. In general, it can be said that the effect of speed errors is to some extent influenced by the geometry and duration of the pass and by the latitude of the observer. For 'high' passes, the longitude error due to east-west speed error can under certain circumstances be greater than the latitude error.

In both parts of the figure the *smaller* component of the fix error (in *Figure 8.11(a)* the latitude and in *(b)* the longitude) depends on whether

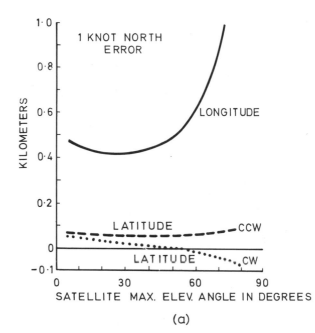

(a)

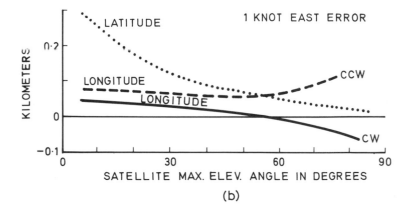

(b)

Figure 8.11 Fix error versus satellite's maximum elevation angle for each knot of error in (a) north speed, (b) east speed. CW = satellite moving clockwise, CCW = satellite moving counter-clockwise (courtesy GSI Texas Instruments)

the satellite travels clockwise (CW) or counter-clockwise (CCW). The CW and CCW curves diverge with increasing elevation angle.

Errors caused by propagation

The propagation of the radio waves from the satellite can be influenced by refraction in the ionosphere (roughly above 80 km) and in the troposphere, the lowest layer of the atmosphere (*Figure 8.12*). The effect of the refraction is, first, that the velocity of propagation changes and, second, that the

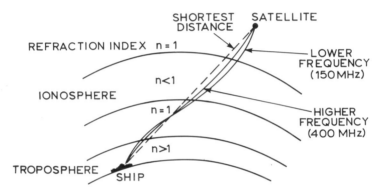

Figure 8.12 The satellite signal's velocity of propagation and path followed are changed by ionospheric and tropospheric refraction. Simultaneous reception of 400 MHz and 150 MHz signals enables the computer to calculate compensating corrections

waves do not follow the shortest path. Both phenomena influence the Doppler count.

Position fixing at night is more accurate than by day, because at night the ionisation and consequently the refraction of the ionosphere are reduced.

The refraction in the ionosphere is inversely proportional to the frequency. That is why the satellite transmits on both 400 and 150 MHz. Both frequencies are received, and by comparing the two signals the computer can calculate the correction to compensate for the influence of the refraction. The refraction in the troposphere cannot be measured in a simple way, however.

The calculation of the refraction is based on a simple model of the ionosphere; this model does not, however, give satisfactory results in the event of solar eruptions; the position fix obtained is less accurate then.

Influence of antenna height

For all position-fixing systems the position of the receiving antenna, not of the receiver, is determined. In other systems the antenna height plays no part, but the satellite system is based upon measuring differences in distance to a satellite far above the surface of the Earth. The height of the antenna above sea-level therefore becomes important.

Since the computer calculations are based upon a reference ellipsoid representing the shape of the Earth, the computer must be provided not only with information about the height of the antenna above sea-level but also with the difference between average sea-level and the reference ellipsoid.

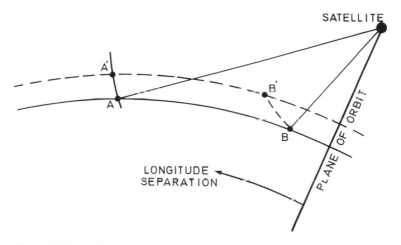

Figure 8.13 Supplying the computer with an erroneous antenna height causes mainly a longitude error

Figure 8.13, showing a satellite passing, makes this clear. The plane of its orbit is perpendicular to the plane of the paper. If the aerials A and B are moved to A′ and B′ their distances to the satellite remain the same. The longitude of A is different from that of A′, however, and for B, nearer to the plane of the satellite orbit, the difference in longitude between B and B′ is even larger.

If an erroneous antenna height is supplied to the computer a longitude error will be introduced; the error increases as the distance from the orbit projection (longitude separation) decreases, i.e. as the maximum elevation angle increases. Course and speed errors also have a great influence then. Therefore *it is not recommended to use passes with a maximum elevation of more than 70°* (see also below).

In *Figure 8.14* the error in nautical miles is shown versus the longitude

separation between the antenna and the satellite orbit (or the corresponding maximum satellite elevation). The curve applies to an antenna height
of about 30 metres; for an exact curve, latitude, geometry of pass, etc.
would have to be taken into consideration.

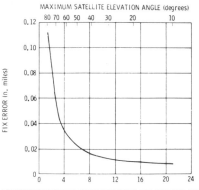

*Figure 8.14 Fix error versus longitude separation
for each 30 metres of antenna height. For example,
if antenna height is 10 metres and longitudinal
separation is 4° (maximum elevation angle about
60°), the fix error is 10/30 × about 0.034 = about
0.01 n. mile*

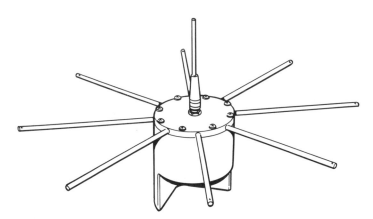

Figure 8.15 Magnavox satellite-receiver antenna, with preamplifier

For special applications Magnavox, an American manufacturer of
satellite receivers, has developed a computer program by which not only
latitude and longitude but also antenna height are determined. *Figure 8.15*
shows a Magnavox antenna.

Inaccuracy caused by apparatus

Variable errors caused by the apparatus can be detected by installing two receivers (with their accessory apparatus) side by side and comparing their readings. Trials have proved that these errors are so small that they may normally be neglected.

Prime satellite passes

A pass with a small elevation angle gives a flat Doppler curve (*Figure 8.8*) and the time at which the distance to the satellite is shortest is not easily determinable. This leads to an inaccurate determination of latitude. Moreover, the tropospheric effect is larger at low elevation, and under those circumstances the refraction correction cannot be calculated very well by the computer.

A more accurate determination of the *latitude* is obtained therefore from passes with a large elevation angle. Such passes, however, do not give a very accurate *longitude* (page 237). The best passes are those

(*a*) that have a maximum elevation angle of more than $10°$ to $15°$ and less than $70°$, and that allow at least four or five counts of two minutes;

(*b*) where an equal number of counts can be performed before and after the centre of the satellite pass.

Such passes are called *prime passes*.

The received frequency is highest as the satellite appears above the horizon, because of its high approach velocity. Every Doppler curve shows this. The receiver is automatically tuned to this high frequency. After the 'time of passing' the received frequency decreases still more but the receiver automatically remains locked to the frequency. After about 15 minutes the satellite disappears again below the horizon. The receiver is then automatically retuned to the higher frequency in order to be prepared for the next satellite.

Sometimes two or three satellites are 'visible' simultaneously. As the receiver can only operate on one signal at a time, some passes will not be tracked at all. If a pass higher than $70°$ or lower than $10°$ (i.e. an unusable pass) begins first, a usable pass between $10°$ and $70°$ may be blocked. It is also possible for the frequencies received from two satellites to become equal. The receiver may then switch from one satellite to the other, thus preventing any fix at all.

To avoid these problems the microprocessor in the Magnavox MX 1102 satellite receiver (see later) determines whether or not a desirable pass is available. If so, the receiver is tuned to acquire the desired satellite regardless of other signals that may be available.

Accuracy of fix at a fixed position

It is difficult to investigate the accuracy of a satellite system at sea because of uncertainties in speed and course information. It is therefore better to do this at anchor or, better still, when moored. A large number of observations can then be made by the ship at this fixed position and, by averaging to obtain the mean fix, accuracy can be considerably increased.

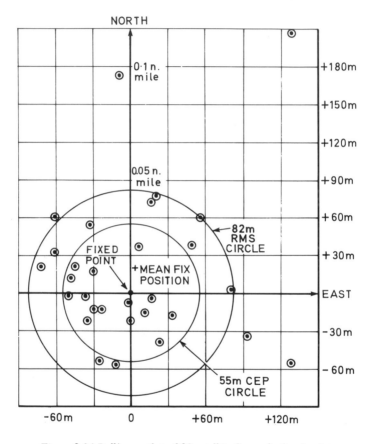

Figure 8.16 Bull's-eye plot of 30 satellite fixes of a fixed point

In *Figure 8.16* thirty fixes taken at a fixed point of known latitude and longitude, and the resulting mean fix, are shown. The r.m.s. error with respect to the true position is only 82 metres. The *circle of equal probability* (CEP) is given a radius such that the number of positions inside and outside the circle are equal. The radius of the CEP in this example is 55 metres.

To improve the accuracy of the fix taken at a fixed point it is desirable to use a well-balanced set of passes to the east and to the west, as well as a number of north-going and south-going satellites. The errors will then partially neutralize each other.

The more passes that have been used, the more accurate the mean fix is. It can be assumed that the mean error of a position fix decreases by $\sqrt{N}$, where N is the number of acceptable positions.

Future improvement of the accuracy of the NNSS system can be obtained by a better knowledge of the position of the satellite as a function of time.

Accuracy of fix on a moving ship

The above procedure can only be carried out when the ship is *stationary* at a known fixed point. According to manufacturers' specifications the accuracy of a position fix taken on a moving ship, using a 'dual channel' (400 and 150 MHz) receiver, is roughly 0.25 n. mile. For 'single channel' (400 MHz only) receivers, the accuracy is 0.5 n. mile.

MAGNAVOX MX1102 SATELLITE NAVIGATOR

Magnavox introduced the MX1102 Satellite Navigator (*Figure 8.17*) in September 1976. During its development mathematical techniques for simplifying the process without degrading performance were found. This simplification of mathematical computations, as well as a carefully coded program to save computation time, enabled the use of a digital microprocessor in the 1102 receiver rather than a larger and more expensive minicomputer. By the use of large-scale integration (LSI) in the microprocessor, the dimensions of the MX1102 (not including the antenna/ preamplifier) were reduced to 420 X 360 X 440 mm. The weight of the unit is 34 kilograms.

The microprocessor shows, among other things, the results of its computations on the alphanumerical video display. Below the c.r.t. is a 'keyboard code' panel, which details the two-digit codes that are most frequently used. Additional codes, less frequently used, can be found by pulling down the keyboard code panel.

The special two-digit codes facilitate initial training and use of the satellite navigator. For example, if the operator wants to know the time the next programmed tracked satellite can be expected, he only needs to use code 51. This is done by pressing keys 5 and 1, followed by E (for enter).

The MX1102 navigates unattended, automatically. After initial data entry, the system can operate indefinitely without operator intervention.

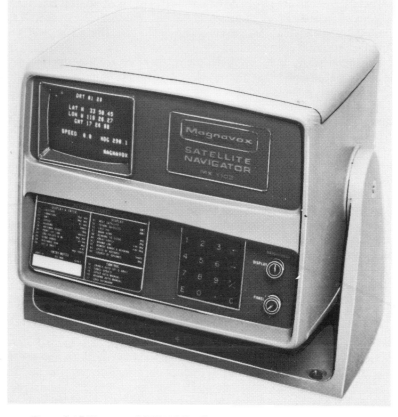

Figure 8.17 Magnavox MX1102 Satellite Navigator (courtesy Magnavox)

It should remain switched on even when in port. To retain total capability when shipboard power is interrupted, batteries automatically maintain system operation for an additional 10–20 minutes. The batteries recharge automatically when shipboard power is restored.

There are only two controls on the panel, one for the brightness of the c.r.t. display and one for illumination of the two-digit codes and keyboard. After the first satellite pass, an internal clock provides a continuous Greenwich Mean Time (GMT) display accurate to one second.

Initialization

When power is first turned on, the MX1102 automatically asks a series of questions necessary for operation. The data required include latitude, longitude, GMT, antenna height, course, speed, and gyro error, if any. Data entry is exceptionally simple through use of the keyboard codes.

Most satellite navigation receivers require the operator to manually enter the geoidal height of the ship at the time of satellite pass. This is not necessary with the MX1102 system, which has been programmed to apply geoidal height correction automatically for all satellite passes.

Automatic dead reckoning

The MX1102 automatically employs dead reckoning to provide position information between satellite fixes (see *Figure 8.18*). An automatic speed and heading interface processes the data directly from the gyrocompass

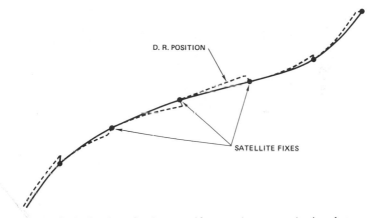

D. R. POSITION

SATELLITE FIXES

Figure 8.18 Dead reckoning provides continuous navigation between satellite fixes

and speed log. Should a gyrocompass repeater or speed log not be available, course and speed can be entered manually via the keyboard.

The dead-reckoning time (DRT) displayed on the c.r.t. is the time since the last satellite fix. This is a measure of navigation accuracy because it indicates how long the system has been dependent entirely on the dead-reckoning data inputs.

Automatic set and drift

The set and drift feature (code 10) of the MX1102 enhances the dead-reckoning (DR) position. If a satellite fix does not correspond to the DR position, the MX1102 is programmed to conclude that a current has carried the ship off the DR track. If enabled by the operator, the processor will compute and apply a set and drift compensation, based upon sequential satellite updates. When the ship is near a region of rapidly changing currents (for example, when crossing the Gulf Stream), manual set and drift estimates can be entered for expected currents. The applied values of set and drift are displayed for easy evaluation as shown in *Figure*

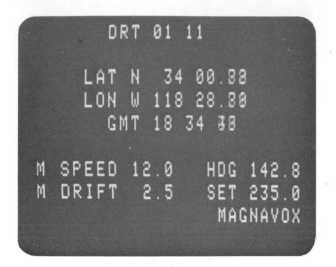

Figure 8.19 Display showing dead-reckoning time, position, GMT, manual (M) speed and drift, heading and set (courtesy Magnavox)

8.19. However, the automatic compensation factor provides improved navigational accuracy in all but a few such situations.

Programmed tracking

Satellite navigation equipment can be adversely affected when two or more satellites are visible simultaneously. *Figure 8.20* plots the satellite passes that occurred in Washington DC on 11 December 1974. These passes have been divided into three categories:

HI Satellites whose maximum elevation was greater than 70 degrees.

OK Satellites whose maximum elevation was between 10 and 70 degrees.

LO Satellites whose maximum elevation was less than 10 degrees.

Satellites that fall into the HI or LO category are likely to have degraded accuracy and are rejected.

When a high or low pass begins before, and overlaps, an acceptable pass,

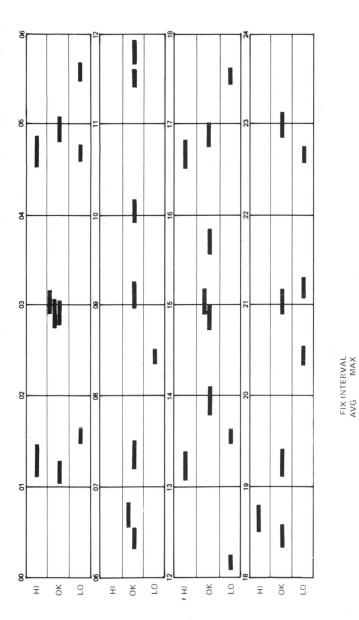

Figure 8.20 Satellite passes on 11 December 1974 at Washington DC (Eastern Standard Time). Microprocessor determines whether an acceptable pass is available, and tunes to that satellite regardless of others available (HI or LO)

a satellite fix may be missed. Furthermore, if the signals from two satellites cross each other in frequency, the receiver has a fifty per cent probability of switching from one satellite to the other, thus denying any fix at all. An example of the first would occur at 1630 in *Figure 8.20*.

To avoid this problem, the MX1102 incorporates a programmed tracking feature; the microprocessor calculates and remembers exactly the orbit of each satellite. On a second-by-second basis, it determines whether or not a desirable satellite pass is available. If so, the receiver is tuned to acquire the desired satellite regardless of other signals that may be available. In the example above, at 1630 the receiver will lock on to and track the HI satellite, whose maximum elevation is above seventy degrees. However, at the moment the OK satellite appears on the horizon, at approximately 1645, the receiver will break track from the HI satellite and lock on to and track the OK satellite. In this way, the MX1102 can provide more and better satellite fixes than systems without programmed tracking. The statistics at the bottom of *Figure 8.20* show that, with programmed tracking computer control, the average time and the maximum time between satellite fixes are reduced.

If the signals from two satellites cross each other in frequency, and the MX1102 follows the wrong one for a time, the program will detect this error and return to the proper satellite signal. However, an automatic position update will not be allowed, thus avoiding a potentially bad fix result. This type of data editing enhances trust in the automatic navigation results.

Short Doppler count

Because satellite signals travel a great distance, a very sensitive receiver is required. Even with a sensitive receiver, sometimes signals fade and are too weak to make possible a reliable Doppler count and fix.

An early improvement in the art of satellite navigation was to increase the number of measurements taken during a satellite pass. By taking 17 measurements of seven seconds each, instead of one measurement of 120 seconds, the likelihood of obtaining a good position fix is greatly improved. This procedure is known as the *short Doppler count*.

Most types of satellite receiver employ a 'majority voting technique', i.e. multiple satellite data are received and a two-out-of-three majority vote is taken to resolve discrepancies. The MX1102 virtually eliminates the possibility of a message error by monitoring the satellite signal strength and simply rejecting data taken during a signal fade. Providing only good data to the majority-vote tests makes the probability of a message error virtually zero.

Automatic inspection

After computation of the satellite fix, and before updating the dead-reckoned position, the fix has to pass inspection. This inspection evaluates the maximum satellite elevation angle, the number of iterations, the update magnitude and the symmetry of the satellite Doppler measurements. 'Symmetry' here means an approximately equal number of Doppler counts before and after the time that the satellite reaches the maximum angle of elevation.

If any of the tests fail (for instance, if the maximum elevation is lower than ten degrees), no update to the DR will be applied. However, in a navigational situation when a borderline update is better than no update, the ship's navigator can override and force the system to update the DR position.

Self-test

To increase further its reliability, every two hours the system automatically self-tests every function. Self-test can also be manually initiated at any time by entering keyboard code 16.

The microprocessor begins by testing itself. First, it executes a sample calculation and verifies the result and, second, it checks the two memories; these tests are followed by others. A relay can be connected to a bridge alarm panel, causing an alarm if the system fails self-test. If a self-test fails, one or more error numbers will be displayed, pointing to the location of the electronic subsystem that failed.

Additional features

In addition to position fixing, the following functions can also be performed:

1. Calculation of range and bearing to a desired way-point (see also page 364). Latitude and longitude of that point must be entered. The navigator can request display of the great-circle or the rhumbline (loxodrome) range and bearing to the way-point. The video display with the result of the calculations is updated every other second to take account of ship's motion.
2. The navigator can request information about the last satellite position fix, including time, latitude and longitude, elevation angle, number of iterations and number of Doppler counts during the passage.

3. The navigator can also request the time of the next satellite pass (between ten and seventy degrees elevation angle) and, if desired, the times of satellite passes after the next one.
4. A printing device can be attached to the MX1102 that will record all the information on the video display either on request or, if desired, every N minutes (N to be selected by the navigator).

COMPARISON OF NNSS SYSTEM WITH OTHER RADIO POSITION-FIXING SYSTEMS

Advantages

1. The system can be used all over the world.
2. The accuracy is greater than that of any other position-fixing system with world coverage.
3. The user does not need to contribute to the enormous cost of the system.
4. Special charts are not required.
5. The position is automatically stated as longitude and latitude, and requires no corrections to be made by the navigator.
6. Lane slip cannot occur.
7. The system does not involve *reflection* of radio waves, so no alterations in propagation time or phase can arise. The refraction in the ionosphere to which the waves are subjected is measured and corrected for by receiving two frequencies simultaneously.
8. The system's computer can also perform other tasks to increase safety or advance efficiency.

Disadvantages

1. The initial cost is high, though this may be reduced in future.
2. The interval (maximum about 4 hours) between two position fixes is long.
3. Errors in the ship's course and speed data supplied to the computer detract from the accuracy of position fixing.
4. It is probable that in the future another, still better, satellite system for position-fixing will become available.
5. The present NNSS system is unsuitable for aircraft because, among other things, the average time between satellite passes is too long relative to the average duration of a flight.
6. Unless the satellite receiver is programmed to select only usable passes, the receiver may perform Doppler counts on passes that are too high or

too low, or may switch over from one satellite to another. In such cases the fixes are inaccurate or useless.

Comparing advantages and disadvantages, the present system is a very accurate and attractive position-fixing system because it lacks some disadvantages inherent to other systems. Nevertheless, there are plans for still better systems of satellite navigation; these systems will not possess the disadvantages mentioned above, and will serve, moreover, for communication.

FUTURE SATELLITE NAVIGATION SYSTEMS

In the United States a new universal satellite positioning system, called the NAVSTAR Global Positioning System (GPS), is in development. Basically it is intended for military use, but will potentially be available for civilian purposes.

The system is not based upon Doppler counts like the Transit system. It will ultimately consist of 24 satellites in circular 12-hour orbits. The height of the satellites is 10 900 n. miles; each satellite completes two revolutions per day. There are three orbit planes, each having an inclination of 63° to the equator plane. In each plane there will be eight satellites; this configuration ensures that at least six of the 24 satellites will always be in view from any point on Earth. On average, nine satellites will be in view, ensuring satellite coverage for three-dimensional positioning.

Each satellite transmits two navigation signals, one at 1575 MHz and the other at 1227 MHz, continuously broadcasting extremely accurate position information. The two signals can be used to determine, and correct for, the effects of ionospheric signal delay.

The ground stations consist of a master and a few monitor stations located in the United States. These stations control and fine-tune the satellites when they pass over each day. As a result of the use of microprocessors with large-scale integration, packed on only two or three printed-circuit cards, the user's equipment consists of a lightweight, small and inexpensive receiver. This equipment can be installed on ships, aircraft and ground vehicles, and even used as a man-pack.

During phase I of the development (the 'test program') four satellites will, in 1978, give a two-hour coverage over the test area each day. By early 1981 (phase II) nine satellites are planned, providing for a two-dimensional capability worldwide. In about 1984 or 1985 (phase III) the 24 satellites will provide the full capability: position accuracy of 7.6 m and 10.8 m in the horizontal and vertical axes, respectively, for 90 per cent of the time, and of 4.0 m and 4.4 m for 50 per cent of the time. (Accuracy figures during phase I will be about twice those for phase III.) Not only position data but also speeds, in three dimensions, will be obtainable with a high degree of accuracy.

9

Radar

GENERAL DESCRIPTION

The word radar is composed of the first letters of *ra*dio *d*etection *a*nd *r*anging. It is an electronic-mechanical aid that not only detects the presence of ships, buoys, the coast, etc., but also measures their bearing and range and often indicates the nature of these objects. In navigation, radar can serve not only for position finding but also as an anti-collision aid. It is evident, then, that radar is a very important aid to navigation, especially since it supplies the same information during the night and in fog as in more favourable conditions. A great advantage of radar is that, unlike other navigational aids, it does not require the cooperation of other stations.

The principle is quite simple: a special transmitter generates very short pulses of radio waves. These are radiated in a narrow beam by means of a directional aerial. When the waves of one of these pulses encounter, for example, another ship, part of their energy is reflected by this ship in all directions – including backwards to the ship from which the pulses were originally transmitted. The reflected pulse constitutes a radio 'echo'. The echo is received by the original ship and, with the aid of a cathode-ray tube (c.r.t.), the time that has elapsed between radiating the pulse and receiving the echo is accurately measured. Since the velocity of propagation of radio waves is known, it is relatively simple to calculate the distance to the other ship. If, for example, the time difference was 100 μs (where 1 μs = 1 microsecond = one millionth of a second), the distance travelled would be 100 $\times$ 300 m, since the velocity of propagation of radio waves is very nearly 300 m per μs. The distance between the two ships would then be half of 30 000 m, which is approximately 8.1 nautical miles. The direction of the other ship is the direction in which the beam was transmitted.

The receiving aerial collects only a very small part of the energy originally radiated, so the transmitter is made to generate very powerful pulses. The transmitting aerial is rotated at constant speed and the beamed pulses are radiated at very short equal intervals, thus permitting the whole horizon to be scanned. The water is also struck by the radiated signals but this radiation is reflected mainly away from the ship.

Range determination by measuring the travel time of radio pulses was first applied in 1926. The purpose then was to measure the heights of ionized layers in the atmosphere. The basic idea was further developed during the years preceding World War II, with a view to military applications. When the war came some British warships were already equipped with what is, by today's standards, primitive radar. It was not until the beginning of the war that, thanks to enormous effort by the British and Americans, radar was quickly developed into a successful system that contributed much to the victory. A major step forward was the development (by Randall and Boot in 1940 at Birmingham University) of the cavity magnetron, which permitted the use of centimetric wavelengths. Thus the German battleship Bismarck and the battlecruiser Scharnhorst, for instance, were finally destroyed by British warships after having been shadowed by radar. It was only after the war was over that the development of radar exclusively for navigational purposes was started.

PRINCIPLES AND OPERATION OF A RADAR INSTALLATION

With the help of *Figure 9.1*, the construction and operation of a radar installation can be briefly described. The block diagram is arranged according to the various functions that have to be performed, rather than according to the physical arrangement of the installation.

It should be appreciated that many types of radar have an arrangement that differs considerably from the simplified block diagram of *Figure 9.1*.

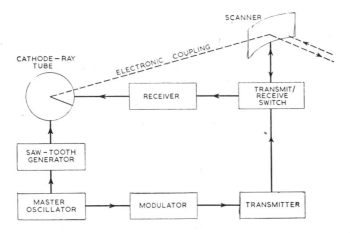

Figure 9.1 Simple block diagram of radar installation

This does not, however, affect this discussion of basic principles. Note also that only shipboard installations will be considered, and not those types used in military equipment and in aeronautical navigation, which must meet other requirements.

Generating the pulses

In the master oscillator electrical pulses of very short duration (between 0.05 and 1.0 µs) are generated. The number of these pulses per second is called the pulse-recurrence or pulse-repetition frequency (p.r.f.). The p.r.f. lies between 500 and 4000 pulses per second. We will assume here a p.r.f.

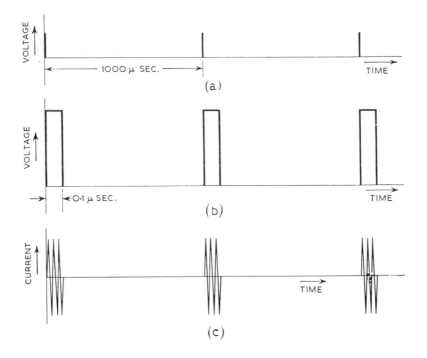

Figure 9.2 (a) Pulses generated by the master oscillator; (b) pulses generated by the modulator; (c) high-frequency oscillations generated by the magnetron

of 1000. These pulses are represented graphically in *Figure 9.2(a)*. Every 1/1000 s (1000 µs) a new pulse follows. The pulses are passed to the modulator whose function is, on being triggered by a pulse, to generate a very high voltage of short duration, say 0.1 µs, and to pass it to the magnetron (*Figure 9.2(b)*).

Transmitting the pulses

The magnetron may be considered as a special type of transmitting valve. It is equipped with a powerful magnet and generates the very strong oscillation of extremely high frequency required in radar. This is done only as long as the modulator supplies a voltage to the magnetron, i.e. each time for 0.1 μs. Considering the dimensions of the magnetron, very powerful signals are generated. This radio energy is supplied to the aerial by way of the waveguide via an electronic transmit/receive switch and is then radiated. *Figure 9.2(c)* shows the radio oscillations graphically.

Note that, in contrast with a normal radio transmitter, the frequency generated by a standard type of cavity magnetron cannot be changed.

In the case chosen here, the radar transmitter generates radio energy for 0.1 μs and remains inactive for 999.9 μs. This period of 999.9 μs is available for receiving a possible echo of the pulse just transmitted.

Receiving the pulses

A sensitive receiver is connected to the same aerial as the transmitter, but to protect the receiver from damage due to the powerful transmitted pulse (power outputs of 30 kW or more are usual), a very fast-working transmit/receive switch is required, one that automatically blocks the receiver during the transmission of a pulse but reconnects the aerial to the receiver immediately after the transmission. This can only be accomplished by an electronic switch, usually known as the T-R switch or T-R cell.

How the picture is generated

The echo produced by the 'target' object is converted into an electric pulse by the receiver, and this is made visible on a cathode-ray tube. In ship radar installations, this is usually done in the following way. The aerial is electrically coupled to coils situated round the neck of the cathode-ray tube, in such a manner that the rotation of the coils is synchronized with that of the aerial.

The windings of the deflecting coils are supplied with a sawtooth current (*Figure 9.3*) from the timebase generator. Each cycle of the sawtooth current must start at exactly the moment that a pulse is transmitted (*a* in *Figure 9.3*), and therefore the timebase generator is triggered by the pulses from the master oscillator shown in *Figure 9.2(a)*. At this instant the c.r.t. electron beam is directed towards the centre of the screen. Because of the increasing sawtooth current the c.r.t. electron beam moves to the edge of the screen and, when the current is maximum (*b* in *Figure 9.3*) the beam has reached the edge. At this stage, however, the negative bias voltage between the c.r.t.'s grid and cathode is such that the electron beam is suppressed; the sweep is therefore imaginary and no light spot is seen on the c.r.t. screen.

Let us suppose for a moment that the aerial is not rotating at the instant at which the pulse is emitted straight ahead and that the coils around the neck of the tube are positioned in such a way that the imaginary light spot moves upwards from the centre of the screen $(0°)$. Shortly

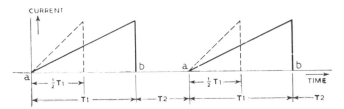

Figure 9.3 Sawtooth current, supplied to deflecting coils, provides radial timebase

after an echo is produced by a reflection, it is received and converted by the receiver into an electric pulse. *This pulse changes the grid bias in such a way that the electron beam is no longer suppressed*. A spot of light on the c.r.t. screen then becomes visible for a moment.

The aerial is now rotated with a constant angular velocity, so that it will perform, say, 15 revolutions per minute. The deflecting coils around the neck of the c.r.t. rotate at the same speed. When the aerial is pointing to starboard, the imaginary light spot moves from the centre to the right-hand side of the screen $(90°)$, and when the aerial points to the stern, the imaginary light spot moves downwards $(180°)$, etc.

The picture on the screen

Figure 9.4(a) represents roughly a ship with radar, a coastline and three surrounding ships A, B and C. As the ship's aerial rotates, these objects are struck consecutively by the outgoing pulses. The echo received from each object will be visible on the screen; the later the echo arrives (i.e. the greater the object's distance from the ship) the further from the centre of the screen will it be seen. (It should be noted here that we shall use the term 'echo' to denote not only the reflected radio signal but also its reproduction on the screen.)

Figure 9.4(b) shows the picture that may be expected on the screen. This image can therefore be considered as a navigation chart of the surroundings, with the addition that objects such as ships etc. appear on it in their correct position. Hence radar can render valuable service as a means of avoiding collisions.

The following calculation will prove that, during the time that transmitted pulse and echo are on their way, the object can have moved only very little even if it is a fast aircraft at a great distance. If the aircraft is at a distance of, say, 50 nautical miles, the echo has to travel 50 nautical miles and this takes approx. 0.00031 s. If the speed of the aircraft is 600 nautical miles per hour (approx. 300 m/s), it will move approximately 300 × 0.00031 = 0.1 m during this time. Thus neither the distance nor

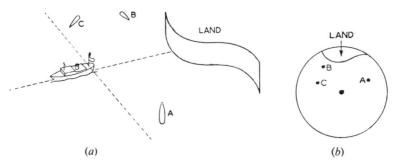

(a)   (b)

Figure 9.4 (a) Ship with radar, three distant ships A, B and C, and coastline; (b) picture on PPI

the direction of the aircraft have substantially changed in the extremely short interval. Also our progress and change in course during 0.00031 s cannot have much influence in this period.

Similarly with the error involved in the rotation of the aerial. At 25 rev/min the change in angle in 1 s is 25 × 360°/60 = 150°, thus in 0.00031 s the change in angle is 0.00031 × 150° = 0.046°, so this error may be neglected too.

We have made it clear that, owing to the extremely high velocity of propagation of radio waves, there is no question of an error in distance or direction when the echoes are reproduced on the screen.

Change of range

Let us suppose that the radar equipment is switched to a six-mile range. That is, an area with a six-mile radius is represented on the screen. An object six miles away will then be reproduced near the edge of the screen. During the time that the outgoing pulse and the returning echo each travel six miles, the light spot on the screen completes one sweep of the radial timebase, and this determines the time T_1 in *Figure 9.3*. The coverage of the radar transmitter is, however, greater than the area shown on the screen. Objects farther away than six miles will also cause echoes, but these will arrive only during the period T_2, when the light spot has fallen

back to the centre of the screen, or during the next period T_1 (this problem will be discussed later). All the echoes that arrive during T_2, irrespective of the direction and range of the reflecting object, will therefore be
reproduced at the same point, namely the centre of the screen. This would
give rise to a troublesome luminous blot in the centre of the tube; to
prevent it, the grid is given, during the time interval T_2, an extra negative
voltage with respect to the cathode. This negative voltage is so high that
the incoming echoes, if any, are not able to override it and set free the
electron beam; such echoes are not, therefore, reproduced.

If the object is three miles away instead of six, the transmitted pulse
and the echo need only half the time to reach the object and return to
the aerial. To reproduce the echo near the edge of the screen, i.e. to
switch over to the three from the six-mile range, the sawtooth current
must be increased to the same maximum value in the time interval $\frac{1}{2}T_1$
(see dotted waveform in *Figure 9.3*). When the range switch is set to
another range the sawtooth current is changed automatically.

The screen

The material of the screen of the cathode-ray tube has the property of
'afterglow'. If this were not the case, an echo on the screen would disappear immediately after its 'birth' and reappear for only a very short
time after another revolution of the aerial. In order to obtain an image of
the whole surroundings, it is therefore necessary to make the afterglow of
the screen material long enough for the echoes to remain visible during at
least one complete revolution of the aerial.

Modern screens are produced with diameters of 9, 12 or 16 inches
(approximately 225, 300 or 400 mm).

MARINE RADAR EQUIPMENT IN MORE DETAIL

The different parts of a radar installation have been outlined with the help
of *Figure 9.1*. A more detailed description will now be given. We have seen
that, when triggered by a pulse, the modulator generates for a very short
time (for instance 0.1 μs) a voltage of about 10 000 V, and supplies it to
the magnetron. The radio-frequency oscillations that the magnetron
generates during this short time are passed to the transmitting aerial, which
takes the form of a short rod inserted in the waveguide. A very strong
alternating electromagnetic field is produced in the waveguide and the r.f.
energy travels along the tube with great velocity. The inside dimensions of
the waveguide are dependent on the wavelength, and its section is usually
rectangular. The dimensions for a wavelength of 3 cm are approximately

12 X 25 mm. Any denting or other damage that affects the inside dimensions of a waveguide will cause loss of power and reduce the efficiency of the radar installation.

The aerial system

In one type of aerial, the horn-like end of the waveguide is located at the focal point of a parabolic reflector (*Figure 9.5*). The extremity is protected by a thin plate in order to prevent the ingress of rain, snow, dirt, etc., because this would have an adverse effect on the operation of the radar installation. It is advisable to brush off the outside of the plate from time

Figure 9.5 Parabolic reflector reflects the radiation emanating from the waveguide. At the end of the horn-like extremity of the waveguide there is a coverplate, which should be cleaned regularly (courtesy Decca Radar Ltd)

to time (say each month) with a dry paint brush, carefully removing salt, soot and other dirt. The dirt absorbs to some extent the energy of the transmitted pulse and the received echo. The coverplate is made of a material that absorbs little energy.

When a source of light is located at the focal point of a parabolic reflector, all the rays form a parallel beam after reflection. In this way, a narrow beam of light can be obtained. In the same manner, i.e. by reflecting the electromagnetic rays emanating from the waveguide extremity with a parabolic reflector, a narrow beam of radiation is obtained with radar. A narrow horizontal beam can only be obtained if the width of the reflecting surface is large in comparison with the wavelength. For a wavelength of 3 cm, which is suitable for marine radar, the reflector dimensions are within practicable limits. *Figure 9.6* shows a reflector that gives a

Figure 9.6 Horizontal beam width of this 3.6 metre radar aerial is only 0.65°

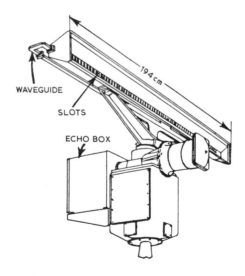

Figure 9.7 Slotted-waveguide radar aerial

very narrow beam of only 0.65°. Note that the fins behind the reflectors in *Figures 9.5* and *9.6* have an aerodynamic function only.

To prevent the horn-like extremity of the waveguide from intercepting too much of the reflected energy, it is located not in the centre of the beam but a little lower.

Figure 9.7 is a sketch of a slotted-waveguide aerial, which has now largely superseded the aerial described above. It consists of a waveguide tube which is supplied with electromagnetic waves at one end, the other end being closed. A number of narrow vertical slots are cut in the tube at the front (horizontal slots are also possible). These slots interrupt the current that is always induced in the inner surface of the waveguide tube, resulting in a narrow beam of radio waves being radiated perpendicular to the front of the waveguide. A simple reflector of small dimensions gives sufficient focusing in the vertical plane.

Some advantages of this type of aerial over the one previously mentioned are that it has less wind resistance, and − for the same dimensions − produces weaker side lobes (see page 304).

In front of the slotted waveguide, a screen of corrosion-proof material is fitted; this does not hinder the radiation but prevents rain etc. from entering the waveguide and reduces still more the wind resistance (*Figure 9.8*).

Figure 9.8 Slotted-waveguide aerial, 3.6 metre span, for 10 cm wavelength (courtesy Decca Radar Ltd)

The rotating portion of the aerial system, by which the directional transmission and reception is effected, is called the *scanner*. It is driven by a motor through gearing which turns the scanner at the required rotational speed.

The horizontal beam width is 0.5 to 2.5° and the vertical beam width is about 20° (see also *Figure 9.34*). The much larger vertical beam width is necessary in order to prevent echoes of objects from being missed as a result of the pitching and rolling of the ship.

Reception

With the type of scanner shown in *Figures 9.5* and *9.6*, the echo signals (which are, of course, exceedingly weak) are received by the parabolic reflector of the aerial and reflected to the extremity of the waveguide where they are propagated downwards to another short waveguide. The T-R switch prevents the echo signals from going to the magnetron, where they would of course be useless. Into the short waveguide is fitted what can be considered as the 'receiving aerial'; this converts the radio waves into electrical oscillations.

The slotted-waveguide type of aerial receives the echo signals directly via the slots.

The radio frequency is so high (≈ 9400 MHz) that it is not practicable in ship's radar to amplify these signals, although it is technically possible nowadays. The frequency is therefore first reduced to a much lower value, e.g. 30, 45 or 60 MHz.

This is done by feeding electrical signals of another frequency into the waveguide branch to which the receiver is connected. This second frequency is generated in a valve of special construction – the *klystron*. Both frequencies produce alternating voltages across a crystal. The frequency generated by the klystron differs by, say, 30 MHz from the frequency of the echo signals, which is that of the magnetron oscillator. As the latter frequency is about 9400 MHz, the klystron frequency will be 9430 or 9370 MHz. As shown in *Figure 9.9(a)*, the two signals together produce beats (see Chapter 1); the number of beats produced per second is equal to the difference of the two frequencies, and in this case is therefore 30 MHz.

A crystal has the property of allowing the current to pass in one direction, while little or no current passes in the other direction. The crystal therefore serves as a detector (see *Figure 9.9(b)*); it consists of a piece of silicon in contact with a wire. *Figure 9.10* is a sketch of a crystal holder.

The current of *Figure 9.9(b)* is smoothed and so converted into a current of the form shown in *Figure 9.9(c)*. The frequency of the resulting alternating current is equal to the number of beats per second, in this case 30 MHz, and is called the intermediate frequency. The received frequency has therefore been converted into a frequency that is sufficiently low for amplification.

It should be understood that the intermediate frequency only exists when an echo signal arrives. The klystron frequency is generated continuously, however, so if there is no echo the klystron oscillations are converted by the detector into a direct voltage, to which the receiver does not react.

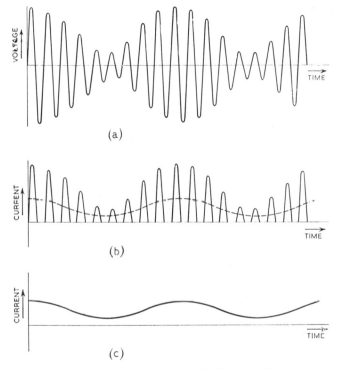

(a)

(b)

(c)

Figure 9.9 (a) Voltage across the crystal when an echo is received; (b) crystal current when an echo is received; (c) current smoothed by capacitor, the number of pulsations being equal to the number of beats at (a)

Figure 9.10 Crystal holder

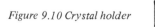

Care should be taken to ensure that spare crystals are always available on board. They are kept in metal capsules to protect them among other things from external electric fields. A crystal should be replaced with caution, carefully following the instructions given by the manufacturer.

The crystal should not be allowed to come into contact with a conductor which is not earthed since this may damage it beyond repair.

To test crystals, most installations are equipped with a meter (test meter) which can indicate the crystal current; the current should lie between two limits indicated by the manufacturer.

Besides spare crystals there are other spare parts on board. Some of them — particularly the magnetron — can seriously influence the magnetic compasses. Therefore they should be stored under normal circumstances at a minimum distance of 10 metres from the magnetic compasses.

Automatic frequency control

In a radar receiver, the amplifier circuits cannot be tuned by hand. This is not necessary since they always have to operate at the same frequency, i.e. the intermediate frequency. Thus the circuits have a fixed tuning.

Now it may happen that the frequency generated in the magnetron or klystron changes slightly. The frequency of the echo signal, and consequently the intermediate frequency, would then be altered. In this case the receiver would no longer be correctly tuned, with the result that the signal is insufficiently amplified. By altering the klystron frequency so that the difference from the magnetron frequency is again 30 MHz, the correct intermediate frequency is obtained and the receiver is once more correctly tuned. In some radar installations this retuning is done automatically, in others manually.

Automatic frequency control (a.f.c.) can be effected in the following way. Part of the energy of the transmitted pulse is diverted from the waveguide through a small gap and mixed with the klystron frequency. Using a second crystal diode a beat frequency is obtained, and this is passed to a special discriminator circuit. This circuit produces a direct voltage that varies according to the value of the beat frequency. This voltage is supplied to the klystron in such a way that it corrects the frequency generated to an extent that will maintain the difference between this frequency and the frequency of the echo signal at about the prescribed value, i.e. 30 MHz. For example, if the magnetron generates a frequency that is 10 MHz too low, then the echo signals also have too low a frequency; a direct voltage is produced and supplied to the klystron, which causes the frequency to be lowered by about 10 MHz. As both frequencies have been reduced by the same amount, the difference between the klystron frequency and the magnetron (and hence echo) frequency remains the same, and in consequence so does the intermediate frequency. In this way the intermediate frequency is always corrected automatically to the value to which the intermediate-frequency amplifier is tuned.

If the automatic frequency control should fail, the klystron frequency can be adjusted by hand in the interior.

A.F.C. is not absolutely essential and some manufacturers therefore dispense with it. In that case the klystron frequency is adjusted manually by a knob that must be set so that the deflection on its associated meter is at its maximum. If this should fail, the klystron should be adjusted mechanically with a socket spanner.

Video amplifier and limiter

It is clear that the intensity of the echoes can vary enormously (ratios of 10^8 may occur). If no appropriate measures were taken, a nearby ship would appear as a troublesome flare spot. At the same time a buoy that is farther away would be scarcely visible. Therefore all echoes are first amplified considerably by the *video amplifier*; their power is increased about 10^{10} times. Then the signals are reduced in the *limiter* to a specific value not exceeding, say, 20 V. Because of the great amplification and subsequent limiting to a fixed value, all objects will be represented by approximately equal light spots.

Suppression of sea-echoes

At sea it will be observed that waves on the surrounding water also cause echoes, though much weaker ones than ship-echoes, etc. This effect is called 'sea-return', 'sea-echoes' or 'sea-clutter'.

Sea-echoes are not seen on the screen as a uniform illumination, but rather in the form of small specks or spots that keep on appearing and disappearing here and there. The echoes of waves close to the ship are naturally stronger than those farther out. This is also due to the fact that at close range the angle of incidence of the radar signals is more favourable to the creation of strong echoes. The amount of sea-return also depends on the condition of the sea surface. The maximum range at which its action is visible is about 3–4 miles. At a closer range the sea-echoes are often so strong that the screen becomes saturated, so that the required echoes from objects such as buoys, etc., are swamped. This is most undesirable since, from a navigational point of view, the echoes from objects close by are very important.

We have just seen that it is the task of the limiter to reduce all echoes to the same level, say 20 V, before passing them to the cathode-ray tube. It is evident that this technique will unfavourably affect the display of, say, a buoy in the presence of sea-return; for when the sea echoes are strong, even though the buoy echo is still stronger, both echo signals will be limited to the same level, and they will appear equally strong. Clearly this is not desirable.

The sea-return may be suppressed by the use of *swept gain*, i.e. an automatic and gradual increase in amplification of each pulse's echoes, from a low level for early echoes to full level for later echoes. Immediately after pulse transmission the receiver applies very little amplification to the echoes; echoes of sea-waves in the immediate vicinity, when amplified, should therefore remain well below the 20 V 'ceiling' applied by the limiter, whereas stronger echoes from better-reflecting objects at the same distance are amplified to a higher voltage (though still not more than 20 V). In this way so much contrast is obtained in the display of the echoes that, for example, a nearby buoy can be distinguished among the sea-return.

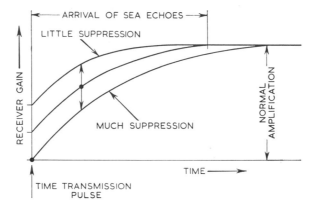

Figure 9.11 Adjustment of sea-return suppression

The receiver gradually increases the amplification of echoes from more distant objects, in such a way that the level of amplification is adapted to the strength of sea-return, which decreases with increasing distance. After the next pulse transmission the amplification is again switched to a low level, and the upward sweep is repeated.

In many types of equipment (see *Figure 9.11*), the degree of sea-return suppression can be adjusted (sea-echo suppression, anti-sea clutter).

Synchro system

The synchro system ensures that the coils around the neck of the cathode-ray tube rotate in synchronization with the scanner. The scanner, driven by a motor, drives in its turn a generator of special construction. In the coils of this generator alternating voltages are induced which are supplied to the synchro-motor fitted in the display unit. While rotating, the axis of the synchro-motor always has the same position as the axis of the generator, so that, but for a slight error, the former follows the latter.

The synchro-motor is coupled by means of a gearing to the deflecting coils. While the coils are rotating, the sawtooth current is supplied to them via two sliprings. A careful check is made that the timebase produced by the sawtooth current in the coils has the same direction as the beam emitted by the scanner.

Range rings

During the time that the light spot, as a result of the sawtooth current, is moving from the centre to the edge of the screen (see *Figure 9.12(a)* and *(b)*), i.e. during the time T_1, a series of pulses is generated at constant intervals by the *calibrator* (see *Figure 9.12(c)*). Like all other video signals these

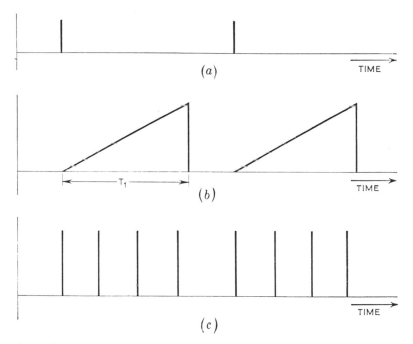

Figure 9.12 (a) Synchronizing pulses; (b) sawtooth current; (c) pulses for range rings

pulses are supplied to the c.r.t., where they momentarily trigger the electron beam, which is normally suppressed during its radial movement to the edge of the screen. If the calibrator produces four pulses at constant intervals, the beam will be released four times (the first pulse, at the centre of the screen, is visible only if the brilliance of the range rings is increased). Three light spots will then appear at equal distances along the timebase. As the timebase rotates, these spots merge into three circles, called *range rings, calibration rings*, or *range markers*.

The time intervals are chosen so that each circle indicates a definite distance on the screen; e.g. on the six-mile range 2, 4 and 6 miles. It is now

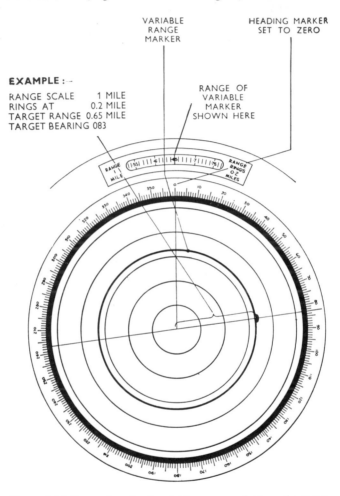

Figure 9.13 Range rings (and variable range marker) on the Decca Radar type TM46

possible to estimate the distance of an echo on the screen. The outer range ring must lie, for Decca radars, about 3 mm within the scale (see *Figure 9.13*).

Although the range rings can be used for estimating the distance, this is not their only function; they also enable the operator to see whether the picture is distorted. If the sawtooth current does not rise linearly, in other words if the oblique line in *Figure 9.12(b)* is not a straight line, the speed of the imaginary light spot along the timebase is not constant. As the time intervals between the calibration pulses are constant, *the distances between the range rings, as measured on the screen, will then become unequal, and the radar picture will be distorted.*

This is illustrated graphically in *Figure 9.14*, where the light-spot deflection (proportional to the sawtooth current) is plotted against time.

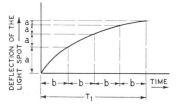

Figure 9.14 Graph of time versus light-spot deflection, showing the effect of a non-linear sawtooth current. (Light-spot deflection replaces sawtooth current as the vertical scale, since the two are proportional.) The time intervals b are equal, but the distances a between the range rings are not

It is not technically difficult to keep the time intervals *b* constant; owing to the non-linearity of the current, however, the distances *a* are unequal, and this is observable on the screen as picture distortion.

The same distortion applies to the entire picture, however, not just the range rings. For example, a ship that is four miles away will still appear on the 4-mile range ring; both echo and ring simply appear a little too far from (or too close to) the centre of the screen. The rings therefore continue to provide an accurate measure of distance, whether distortion is present or not. This applies not only to the fixed rings but also to the variable range marker described later.

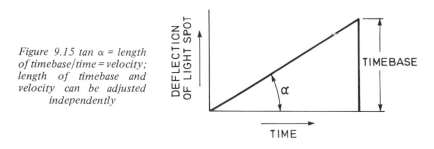

Figure 9.15 tan α = length of timebase/time = velocity; length of timebase and velocity can be adjusted independently

In most radars, there are variable resistances (potentiometers) by means of which the sawtooth current can, if necessary, be made more linear; the resistance controls are marked 'linearity'. The exact adjustment is the one where the distances *a* of *Figure 9.14* are equal.

There are other variable resistances by means of which the length of the timebase can be varied. These resistance controls are marked with the word 'length'. After adjusting the linearity the length must be readjusted.

Figure 9.15 shows the ideal sawtooth current, with time horizontally and the deflection (proportional to the sawtooth current) vertically. The tangent of angle α = the length of the timebase divided by the time = the velocity of the light spot. By means of a third control this velocity, i.e. the magnitude of angle α, can be altered. It is obvious that by doing this the distances between the range rings can be increased or decreased.

The display unit is provided with a knob by which the range ring pulses can be made stronger or weaker, and in this way the brilliance of the rings can be varied.

Variable range marker

Nearly all radars can also be equipped with a variable range marker (travelling or moving range ring). This is a range ring the diameter of which can be varied with the aid of a control. In this way the operator can adjust the ring's radius until its outer edge touches the nearest point of a selected echo. A pointer on a scale beside the screen then indicates the distance to the echo directly in nautical miles and fractions of a mile (*Figure 9.13*). In some installations the variable marker's radius is indicated by a digital counter instead of a scale and pointer.

To produce this variable range marker, the synchronizing pulse from the master oscillator is supplied to a variable range pulse generator, which then generates a single pulse. The instant at which this pulse is generated after receiving the synchronizing pulse is varied by the control, and thus the radius of the marker increases or decreases. Non-linearity of the sawtooth current affects the picture and the fixed and variable markers equally.

It is desirable to test the accuracy of the variable range marker daily. (The fixed markers generally maintain their accuracy longer than the variable one.) By matching the variable range marker with the smallest and the largest of the fixed range rings, the calibration of the variable marker can be checked. If in both cases the reading on the distance scale is correct, it may be assumed that it tallies for intermediate values also.

If the reading is not correct, the control for the variable range marker is turned until the correct distance is shown on the scale. As a result, the fixed and variable rings will no longer coincide. Expert personnel can now adjust a variable resistor, often marked 'range trimmer', mounted in the interior of the display unit. This changes the radius of the variable marker, which grows smaller or larger until it matches exactly the fixed range ring, but does not affect the reading on the scale of the variable marker. This procedure is followed first for the smallest and then for the largest ring in each range, after which a new check and possibly further readjustment is required.

Performance monitor

A not essential but very desirable component is the echo box, which is mounted in the vicinity of the scanner, and which forms part of the non-rotating part of the aerial system. It consists of a box having a cavity with one aperture (see *Figure 9.16*). Each time the rotating aerial radiates pulses in the direction of the echo box, part of the energy is received

Figure 9.16 Slotted-waveguide aerial fitted with echo box

through the aperture. The dimensions of this box are arranged to bring it into electrical resonance with the radar frequency. Owing to various causes the echo box may be out of resonance. A small rod is therefore fitted which is set spinning quickly by a small motor. This changes the echo cavity dimensions, which are important from an electrical point of view, so that there is resonance for definite positions of the rod.

The electromagnetic oscillations induced by the radar signal gradually die out in the cavity and are radiated back through the aperture to the scanner. Then the picture shown in *Figure 9.17* appears on the screen, resembling the plume of a feather. The stronger the pulses are, the longer it takes for the echoes from the box to die away and the longer the plume will be on the screen. The plume will also become longer when the intermediate frequency is such that it is in resonance for the circuits of the receiver. As already described the intermediate frequency can be changed with the aid of the klystron.

With normal amplification and brilliance adjustment, the plume should have a specific prescribed length, which is approximately the original length when first adjusted on setting up the installation. This length may correspond to, say, one nautical mile. It is important that operators know this length or are able to find it (e.g. in the radar log). If the length is

Figure 9.17 Plume caused by reradiation from echo box

shorter than it should be and if the error is, as usual, not due to the echo box, then it should be corrected. The cause may be reduction of the emitted power, incorrect adjustment of the klystron frequency, etc.

The echo box thus gives a positive indication of the satisfactory operation of the entire installation, and functions as a performance monitor.

A monitor that causes a plume should be adjusted so that the plume lies in the direction of a blind sector, if any (see page 309).

Sea-echo suppression must be suspended during checks of the radar installation by means of the monitor, otherwise the receiver gain for short distances would be reduced too much. On the other hand, the suppression of sea-echoes can be checked by the suppression of the plume of the echo box to a satisfactory extent.

In some radar installations the transmitter and receiver can be checked separately by the TX monitor and the RX monitor. The TX monitor is a neon valve that replaces the echo box and is struck by the aerial beam at each aerial rotation. The neon gas is thus ionized and this produces on the screen a plume which is longer as the transmitter generates stronger pulses. For the RX monitor, a small quantity of energy is withdrawn from each transmitted pulse. This energy is supplied to the receiver and causes a circle to appear on the screen. Depending on the amplification of the receiver, the radius of the circle becomes larger or smaller.

MAIN UNITS

The components of a typical marine radar installation are grouped into four main units (*Figure 9.18*):
1. The motor-generator with its voltage regulator, which may be mounted in any suitable place. The motor is connected to the ship's mains and drives the alternator that generates the alternating voltage required for the radar installation. The voltage regulator keeps the voltage generated by the alternator constant, despite fluctuations in the mains supply.
2. The transmitter unit and the receiver (together called the transmitter/ receiver or transceiver). This unit comprises, for instance, the modulator, the magnetron, part of the receiver or the entire receiver. There are types where the transmitter is contained, for instance, in the aerial system or the receiver in the display unit. The unit is usually situated in an easily accessible and protected spot as near as possible to the aerial.
3. The display unit containing the cathode-ray tube, etc. The operating controls (e.g. see *Figures 9.19* and *9.20*) are located here. The unit may be mounted in the wheelhouse, the chart room or (the trend in new ships) in a special room that can be kept dark and where there are at the same time facilities for plotting. It is desirable that the radar operator observing the display should also have, as far as possible, a clear view straight ahead.

A remote display unit may be installed in another part of the ship; this slave unit need not be identical to the master unit (e.g. it may lack some controls). A second possibility is to install two identical radars, each with its own aerial, etc.; this allows interchange in the event of faults; see *Figure 9.18(b)*. A third possibility is to have two *different* radars (one with a wavelength of 3 cm, the other 10 cm, say). One display unit can then be switched to a short range, the other to a long range; thus the navigator can be warned at an early stage of the

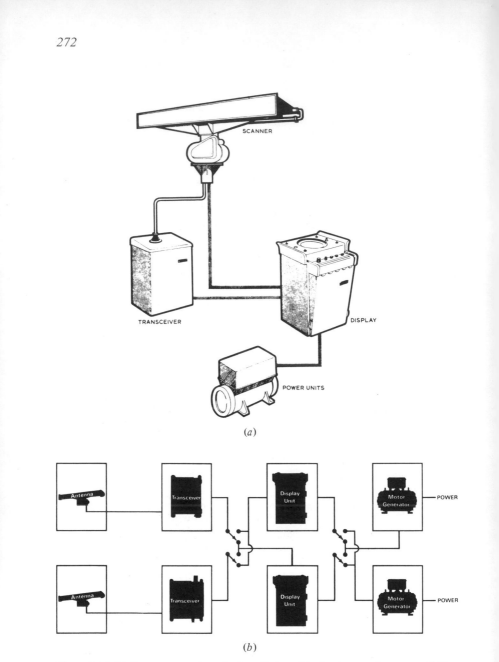

(a)

(b)

Figure 9.18 (a) Components of a radar installation; (b) where two radars are available, their display units and power supplies can be interchanged

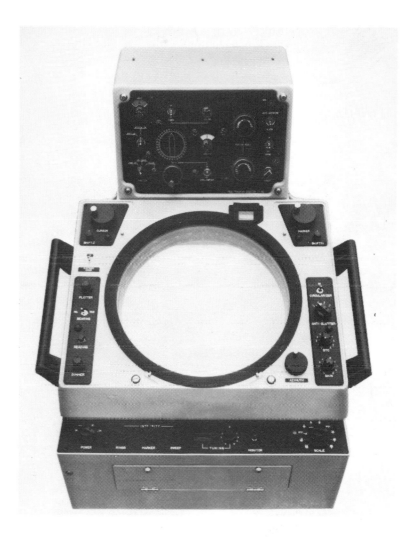

Figure 9.19 Sperry Mark 16 display unit. At the far side of the screen is the 'true tracking adaptor'. Controls on both sides of the screen: cursor, shift, electronic cursor, plotter, bearing (rel/true), heading, dimmer, marker, shift, circularizer, anti-clutter, STC, gain. Controls on the near side of the screen: power, intensity of rings/marker/sweep, tuning, monitor, scale (courtesy Sperry Rand)

approach of distant ships. An attractive and practicable solution is to switch one display to 'true motion' (see later) and the other to 'relative motion'.

4. The aerial system, which chiefly consists of the slotted-waveguide or reflector assembly, the driving motor and the synchro-generator. Sometimes the transmitter and part of the receiver are mounted in the aerial unit. The waveguide is then much shorter, thus reducing losses, but

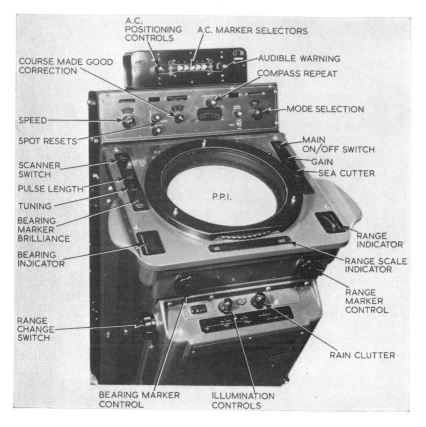

Figure 9.20 Decca 66AC display unit (courtesy Decca Radar Ltd)

with this arrangement the problem arises of keeping the components dry while repairing them, especially in bad weather; sometimes the entire unit may have to be removed bodily from the aerial system for servicing or repairs.

The presence of water or moisture in the waveguides, etc., reduces the radiation materially, if not completely, and it is therefore common to install electrical heating equipment. A thermostat may sometimes be used to keep the temperature of the apparatus at a constant level. To facilitate the repair or inspection of the aerial, a telephone connected to the display unit is sometimes installed near the aerial. A switch may also be located there for switching on and off the driving motor of the aerial.

The scanner should obviously be mounted as clear as possible of the obstacles present on board ship in order to be able to scan the surrounding area fully. The radiation straight forward should not be interrupted in any way that causes a blind sector (see page 309). A high mounting increases the radar's range, but, owing to the unfavourable angle of incidence, makes short-range sea-echoes more pronounced (see page 263), and this is not desirable.

When the radar installation is switched on while the ship is in port, care should be taken to ensure that the reflector can rotate freely. If this should be hindered by steel wires or rigging the driving motor can sustain severe damage. Some types have a separate aerial switch.

OPERATION OF THE RADAR

In order to obtain optimum results, it is necessary, of course, for the whole installation to be tuned and adjusted in the right way. The switches and controls fitted for these purposes may be classified into two groups.
1. Those that are often used by the operator are mounted within his immediate reach on the display unit.
2. Those that are less frequently used by the operator are also located on the display unit, but with some types they are fitted underneath a lid.

In addition to these two groups there are the preset controls, which are in the interior of the display unit, modulator-transmitter and aerial system and mostly are only accessible after removal of a plate. These have been adjusted after installation by expert technicians. The adjustments are frequently made with a screwdriver; in course of time they should be rechecked and, if necessary, altered by technicians, even if the installation appears to be still in good working order.

Fuses are usually found behind a front plate or other plate or lid; in many types a blown fuse may be detected by the lighting up of a small neon valve mounted near it.

The symbols for the various radar controls, as recommended by the IMCO, are listed on pages 351 and 352.

Power switch

On the display unit we find first the power switch with which the motor-generator is brought into operation. Once the motor is started, power is supplied to the filaments of the cathode-ray tube(s) (and valves, if any), which are gradually heated. In modern radars, most valves (but not, of course, c.r.t.s) have been replaced by semiconductor devices.

In addition to the filament voltage, c.r.t.s and valves also require the much higher anode voltage. Only when the cathodes have reached their required temperature, i.e. after about 1−2 min, may the anode voltage be applied; if not, damage would be caused. Therefore the high anode voltage is automatically switched on after the appropriate warming-up period by means of a multi-contact relay, after which the instrument is ready to operate.

Often the main switch has a third position: 'stand-by'. In this position, the instrument is ready for immediate use but does not yet function; the filament of the c.r.t. (and any valves) is heated but the anode voltage is disconnected. In this position there is no radiation or timebase. The power consumption is lower, of course. The lifetime of the installation can be increased by switching from the 'on' position to 'stand-by' since frequent switching on and off is detrimental to c.r.t.s and valves. If, therefore, the radar is not required for a short time but may be required suddenly, one should not switch it off but turn the switch, if possible, to the 'stand-by' position.

In the 'off' position, with many installations, heating resistances in aerial system and display unit remain switched on in order to keep the temperature as constant as possible. During a longer stay in port they are usually switched off, a special switch being provided for this purpose.

Many types are provided with an hour meter or time-recording meter, which records the number of hours that the instrument has been working (excluding stand-by hours).

Focus

As remarked before (page 27), it is necessary for the electrons to converge as much as possible on the screen, so that the light spot has the smallest dimensions and the picture is formed by thinner lines and is consequently sharper. Usually this is obtained by giving the direct current in the focusing coil the correct value with the aid of a variable resistance. The control concerned is marked 'Focus' (see *Figure 9.21*).

If the adjustment is correct at the centre of the screen, it may nevertheless sometimes be less accurate towards the edge of the screen. It is best, therefore, to adjust for sharp focus on a range-ring half-way between the centre and the edge.

It may be that when the equipment is first switched on, or is switched to another range, the focusing will need some readjustment. There are types where the focus is preset and requires screwdriver adjustments.

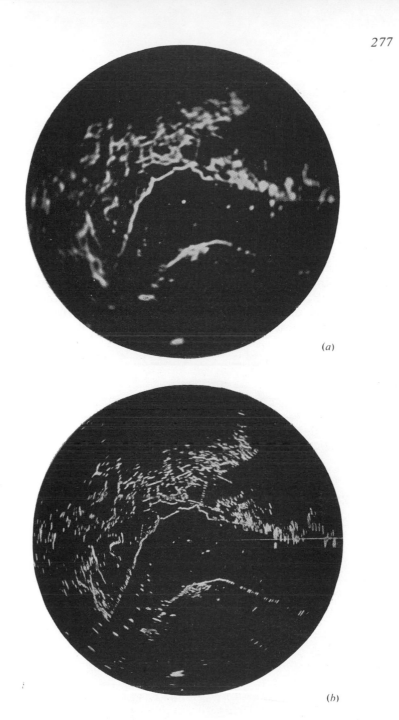

(a)

(b)

Figure 9.21 Focus: (a) incorrect; (b) correct (courtesy Decca Radar Ltd)

Brilliance

The brightness of the display may be regulated by a control marked 'Brilliance'. This control adjusts the direct voltage between the grid and the cathode of the c.r.t. The more negative the grid is relative to the cathode, the weaker will become the electron beam until it is eventually extinguished. The brilliance of the whole screen, range rings and markers included, will thus depend on the setting of this control. This is shown in *Figure 9.22. The correct brilliance is obtained when the rotating timebase becomes just invisible*. In practice the control is first turned up and then turned back until the timebase just vanishes.

In order to avoid interference from other light spots (the picture itself, range rings, etc) while adjusting the brilliance control, it is advisable to keep the receiver gain control in the minimum position.

A primary result of excessive brilliance (see *Figure 9.22(c)*) is that (in the case of normal amplification) the echoes in the picture do not stand out with sufficient contrast to the background. Secondly the life of the c.r.t. is reduced, and thirdly the focusing may be affected. If the brilliance control is turned back too far, the desired echoes will also gradually disappear (see *Figure 9.22(a)*).

(a)

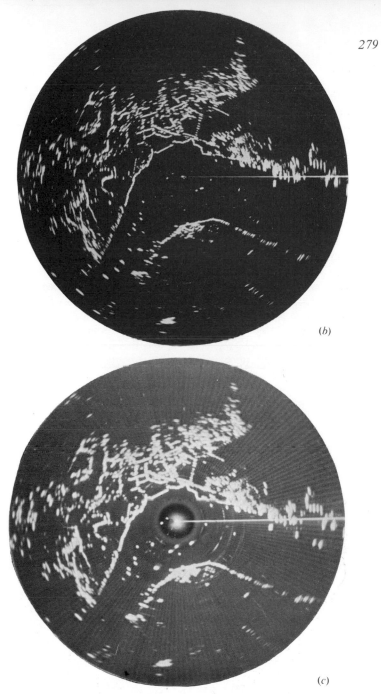

(b)

(c)

*Figure 9.22 Brilliance: (a) too little; (b) correct; (c) too much
(courtesy Decca Radar Ltd)*

It is advisable to readjust the brilliance from time to time, and also when the set is switched to another range.

Although the rotating timebase should be suppressed, it may nevertheless be detectable because of the echoes, the 'grass' and the faint light it leaves behind on the screen. With some types, chiefly British ones, the brilliance, and sometimes the amplification too, should be turned down before switching off the installation and turned up again after the installation has been switched on by the time relay. If this is omitted, the screen may be burnt in the centre.

Receiver gain

After amplification in the receiver the incoming echoes are passed to the c.r.t. The negative grid voltage is reduced by this to such an extent that the electron beam is no longer suppressed. The correct receiver gain or, which is the same, the exact sensitivity must be adjusted with the utmost care with the control marked 'Receiver gain', 'I.F. gain' or 'Sensitivity'.

As previously stated, 'noise' will occur at all frequencies just as in radio receivers, and little can be done to prevent this interference. It cannot be completely suppressed since it is largely generated in the early stages of the receiver at all frequencies, and is thus amplified along with the signal frequency. In normal radio receivers possessing sufficient amplification, this form of interference is audible as a continuous hiss or noise, but in a radar installation it becomes visible as a faint and irregular illumination over the entire screen. This is called 'grass'. Of course the separate spots of the grass do not always appear in the same place on the screen.

As the gain of the receiver is increased, the grass becomes brighter and more extensive until at length it is impossible to distinguish the real display. The grass disappears completely when the gain is reduced sufficiently, but at the same time the weaker echoes may then be insufficiently amplified; it might happen that a weak but important echo could disappear altogether. The gain should therefore be increased as much as possible without the grass becoming too troublesome. The correct gain is reached when the grass shows as a faint speckled background in which all the details of the picture may be clearly distinguished (*Figure 9.23(a)*). *Figure 9.23(b)* shows the same picture with excessive gain. Periodic adjustment of the gain control is even more important than adjustment of the brilliance control.

Video gain

In some radars a control marked 'Video gain' is provided, usually of the preset type; it controls the amount of amplification applied by the video

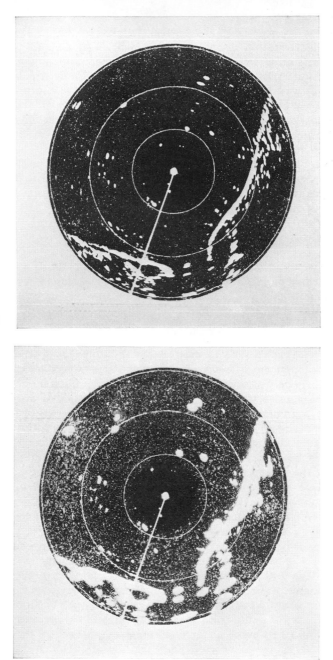

(a)

(b)

Figure 9.23 Receiver gain: (a) correct, with grass visible as a speckled background; (b) excessive (courtesy Sperry Rand)

amplifier (see page 263). If the signals supplied by the video amplifier (via the limiter) are too strong, the electron beam in the c.r.t. has the same effect as a jet of water striking a sheet of water; namely that other drops of water are splashed up and then fall back again in the neighbourhood. In the same way too strong an electron beam striking the screen on receipt of echoes will make other electrons eject from the screen and fall back in the neighbourhood. Instead of a small spot of light, a large illuminated area will then be visible. This is called 'blooming'. (It will be clear that this phenomenon has nothing to do with focusing.)

When the 'Video gain' control is correctly set, the signal and other voltages supplied to the limiter, and thence to the c.r.t., are not high enough for blooming to occur, even when the 'Receiver gain' control is set to its maximum.

Sea-echo suppression

Sea-echo suppression (sometimes termed 'anti-sea clutter') was previously discussed on page 263. Before using this control we should first try slightly reducing receiver gain; a small reduction in gain may prove helpful. However, there are many occasions when sea-echo suppression will be needed. Without suppression, the echoes from small buoys could be missed. When adjusted correctly, sea-return can be suppressed so that buoys will just be visible. Sea-echoes show as small specks and dots that continually appear and disappear in different places on the screen. Echoes of buoys, of course, always appear in the same spot. The very instability of sea-echoes and the constancy of a buoy-echo provide a means of distinguishing a buoy even in the presence of appreciable sea-return.

It is of course possible to adjust the suppressor control too far, which may cause the loss of desired nearby echoes owing to their amplification being reduced. If, on the other hand, the degree of suppression is insufficient, the desired echoes will be swamped in the sea-return. The control should be set, therefore, so that the sea-echoes are only just visible as small, not too troublesome, specks. When the display of the surroundings of the ship is congested and confused as a result of nearby buildings, etc., the use of sea-echo suppression will help clear the display.

Particular care should be taken that small craft do not approach unnoticed as a result of excessive suppression. The echoes from small craft are unlikely to be detectable until they are at fairly close range, just outside the area of sea-return; excessive suppression (*a*) extends suppression to the area where such craft can first be detected, and (*b*) suppresses small-craft echoes, together with sea-echoes, within the sea-return area itself.

Figure 9.24 consists of three photographs which show the pictures on the screen of the Shorebased Radar Station at IJmuiden. In (*a*), no sea-echo suppression has been applied; ships inside the piers would go unnoticed owing to the heavy sea-return. In (*b*) the suppression is so much

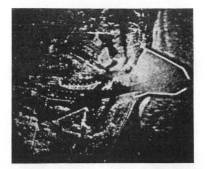

(a)

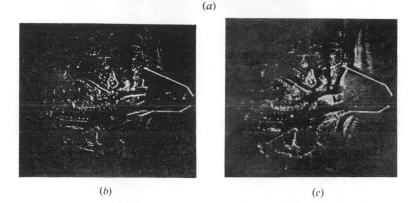

(b)   (c)

Figure 9.24 Sea-echo suppression on a radar display at Philips' Shorebased Radar Station at IJmuiden: (a) no suppression; (b) maximum suppression — sea-return has disappeared completely, but so have nearby echoes; (c) correct suppression — objects at close range are clearly visible in the sea-return (courtesy Philips Telecommunication Industries)

that the echoes from small craft have probably also disappeared. This setting is, therefore, dangerous. In (c) the correct degree of suppression has been applied. Some objects at close range are still clearly visible in the sea-return.

Anti-clutter rain/snow

Some radar sets include a two-position switch labelled 'off' and 'rain/snow'. Heavy rain and snow create a more or less evenly illuminated picture; echoes from ships or coastlines can no longer be clearly observed. The anti-clutter rain/snow switch, which is normally in the 'off' position, can then be turned to the 'rain/snow' position with the result that the desired echoes are better contrasted against the snow and rain echoes (see *Figure 9.25*).

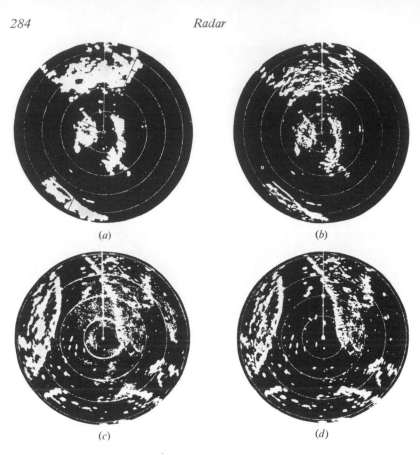

(a)   (b)

(c)   (d)

*Figure 9.25 Anti-clutter rain/snow: (a) display with rain echo; (b) same display
with rain-echo suppression and, to a lesser degree, sea-echo suppression; (c)
display of Narrows, New York Harbour, with rain- and sea-echo; (d) same display
after suppression of rain- and sea-echo (courtesy Sperry Rand)*

The method used to eliminate the rain/snow clutter around the echo
from a ship or from some other object is based on the fact that the echoes
of all the drops of rain, which together give the impression of a compact
mass, are of similar (fairly low) strength. Echoes from objects like ships
are normally stronger, owing to better reflection (see *Figure 9.26(a)*). A
circuit called a *differentiator* reacts to sudden changes in voltage, and
accentuates them; supplied with a voltage pattern like that in *Figure 9.26
(a)*, a differentiator therefore produces the pattern shown in *(b)*, and on
the screen the strong (ship's) echo will be clearly shown against a faint
rain-echo background.

A differentiator circuit contains a capacitor and a resistor, and its 'time
constant' is the product of the capacitance and resistance values. For our

purpose the time constant is required to be small, or 'fast'. Hence the rain/ snow suppression switch is often labelled 'FTC' (fast time constant).

In heavy precipitation, the radiated energy is absorbed so much that the range is reduced. Behind an area of heavy rainfall, a shadow will therefore appear to a greater or lesser degree. If ships are to be detected *behind* areas of rain, more amplification may sometimes help. Where objects are to be detected *within* areas of rain showers, improvement is effected by *reducing* the amplification.

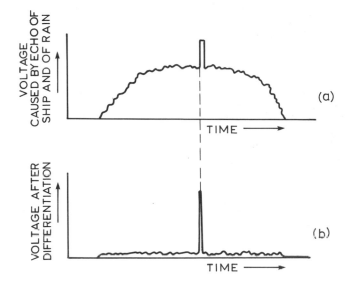

Figure 9.26 (a) Ship echo is usually stronger than rain/snow echoes; (b) differentiator produces a pulse at the beginning of the arrival of a strong echo

If the radar wavelength is 10 cm, less trouble is caused by rain and snow echoes on the screen than with 3 cm equipment.

The anti-clutter rain switch may also provide a less confused picture where there is congestion as a result of echoes from buildings, etc.

Switch for choosing pulse length

One widely used radar design has a control with which either of two alternative pulse lengths may be chosen. When switched to 'long', the pulse duration is 1 μs and when switched to 'short' it is 0.1 μs. The longer the pulse, the greater the energy of the transmitted pulse and its echo. How- ever, with the longer pulse the echoes of two objects close together on the same bearing will tend to merge into one spot of light on the screen. It is then more difficult, for example, to discern an object in the sea-return. This subject is discussed again under 'range discrimination'.

Test meter

Many radar sets are equipped with a test meter and ancillary switch. For each position of the switch, the meter indicates the current or voltage in a specific circuit. Comparing the meter readings with the data supplied by the manufacturer affords a quick and easy check of the equipment. The meter can also be used to trace possible faults. On installations without automatic frequency control, the meter is used in setting up the correct tuning adjustment. The switch on these sets should be turned to a position marked 'tune', and the control associated with the tuning must be turned until the meter needle shows maximum deflection. This control alters the frequency by varying the direct voltage applied to the klystron oscillator.

Range switch

The 'Range switch' or 'Range selector' is an important control by which the radius of the displayed area may be varied. A choice of seven fixed ranges (¾, 1½, 3, 6, 12, 24 and 48 nautical miles, ratio 1:2 at each step) is offered with some makes. The smallest range normally found is ¼ and the largest 64 nautical miles. The range to which the set is switched is clearly indicated on every type of display unit.

Range rings

Some units are equipped with a range marker control by which the brilliance of the range rings or markers may be varied from invisible to an intensity equal to the strongest echo. It is inadvisable to increase the intensity more than is strictly necessary, although the danger of overlooking an echo just covered by a range ring is actually not great.

Variable range marker

Nearly all makes are fitted with the variable range marker previously described, and with a scale or counter to indicate the distance corresponding to the radius of the marker.

Heading flash

A control marked 'Heading flash' or 'Heading marker' is fitted to all radar sets. With this, a line of light can be made visible from the screen centre upwards to $0°$. This indicates on the screen the ship's centreline in the direction straight ahead. Its purpose is to enable the operator to tell at a glance whether objects judged to be approximately straight ahead are located to port or starboard. This is, of course, very important in navigation.

There is a danger of a weak echo dead ahead being covered by the heading flash and consequently not being noticed; yet this danger is not as great as would be presumed, and it is generally advisable the keep the ship's heading flash switched on in normal circumstances.

Because of cross-set, the heading marker does not necessarily indicate the direction of the ship's movement with respect to the sea-bottom, as the appearance of the radar display might suggest.

The heading marker is produced by a switch in the aerial system, the scanner microswitch being closed or interrupted for a short while during the rotation of the reflector at the moment when the scanner is radiating in the direction of the ship's bow. The closing or interruption of this switch indirectly releases the electron beam in the c.r.t. so that the time-base (or rather, a few timebases, together constituting the heading flash) becomes visible.

It is advisable to check from time to time that the microswitch is closed or interrupted at the correct position of the aerial, and also whether the other circuits concerned are properly adjusted. If this is not the case, the heading flash will be pointing in the direction $0°$ whilst the scanner is not radiating at $0°$. The check can be made when an object that gives a good reflection but is not too large (for instance another ship) is seen dead ahead from the bridge. The heading marker should then bisect the echo from the ship. If this is not the case it will be necessary to adjust the scanner microswitch so that the heading flash turns. When the microswitch is corrected and the heading flash is in line with the ship, it may be that the heading flash is no longer pointing in the direction $0°$. There is now a difference in angle between aerial and picture including heading flash. In the usual way (see page 299) the picture must then be turned so that the heading flash points to $0°$.

Cursor

The bearings of echoes on the screen can be determined with a rotating cursor. This takes the form of a perspex disc, with a diameter etched into both front and back, mounted on top of the screen. The eye should be positioned so that the two lines coincide, thus avoiding the parallax error. The disc sometimes has two mutually perpendicular diameters etched on it; each of the lines can serve as bearing cursor. A control is fitted that revolves the whole disc mechanically. A scale graduation of $0°-360°$ is fitted round the circumference of the screen, $0°$ being directly above the centre (see *Figure 9.13*).

The centre of rotation of the disc must of course coincide with the centre of the scale. A regular check should also be made to ensure that the mechanical centre of the disc coincides with the electronic centre of the display, i.e. the axis of rotation of the timebase. If this is not the case, errors will arise in the bearings, especially those of echoes near the centre of the screen.

When a bearing is taken, the cursor line should bisect the echo exactly.

Further lines parallel to the diameter may also be etched on the disc (*Figure 9.30*). If these are turned parallel to the heading marker and a fixed echo observed, the presence of 'cross-set' can be detected. This is the case if the trace of the echo is not parallel to these lines.

Recent radars are also fitted with an *electronic cursor*. This cursor is obtained by not suppressing the timebase in a particular direction, which results in a radial line of light appearing on the screen. This line can serve as a cursor. Its position can be adjusted by means of a control. For distinguishing purposes we will call the cursor described first the *mechanical cursor*.

It will be seen later that the electronic cursor is essential for 'true motion' displays. The bearing is then read from a special scale, which is not placed round the edge of the screen.

One advantage of the electronic cursor is immediately apparent. There cannot be any parallax error. A second advantage is that any distortion will influence the cursor and the picture in the same way, which causes cancellation of possible errors.

We have seen that, when using the mechanical cursor, errors in the bearing arise if the electronic and the mechanical centre do not coincide. These errors can be avoided if the electronic cursor is used and the bearing is read from the special scale. So this is a third advantage.

A disadvantage, however, is that the aerial must make one revolution between each 'repainting' of the electronic cursor. In the meantime, the brilliance of the cursor gradually fades but its position does not change, even when the appropriate control is turned. The bisecting of the echo is thus rendered somewhat more difficult.

A second disadvantage of the electronic cursor is the danger of confusion with the heading marker. One manufacturer has solved this by letting the cursor begin not from the centre but from the variable range marker. On the equipment made by another manufacturer the cursor is interrupted at regular intervals, and so appears as a broken line. If there are no such differences between the electronic cursor and the heading marker, it is possible to adjust them to different light intensities to prevent confusion.

Interscan

Another and technically better method for obtaining an electronic cursor is the interscan method.

For this purpose, besides the normal deflecting coils, extra 'interscan' coils are arranged around the neck of the cathode-ray tube. During the periods in which no sawtooth current passes through the normal deflecting coils (in *Figure 9.3* the time intervals T_2) a sawtooth current that has been generated separately is supplied to the interscan coils. This causes an extra timebase perpendicular to the axis of the interscan coils. The latter can be rotated to any position by means of a control marked 'Interscan bearing marker'. The extra timebase makes the same rotation and thus can serve as cursor.

With 'true motion' the axis of rotation of the timebase does not coincide with the centre of the scale. Therefore the bearing scale round the edge of the screen cannot be used and the position of the interscan bearing marker must be read from a special scale.

At long ranges, the time interval T_2 in *Figure 9.3* is too short. Every 20th (or 30th) normal sawtooth oscillation is then supplied to the interscan coils instead of to the rotating deflecting coils. In the case of a pulse-repetition frequency of, for example, 1000, the same number of sawtooth oscillations per second are generated, and 1000/20 = 50 are supplied to the interscan coils. This is sufficient to avoid a flickering cursor.

Decca radar has provision for altering the length of the interscan bearing marker with two controls, the 'Interscan range coarse' and the 'Interscan range fine'. The distance in nautical miles corresponding to this length is indicated on a distance scale. In this way the bearing marker can be used both for taking bearings and for measuring distance.

The brilliance of the cursor can be adjusted.

Dimmer

The bearing scale, distance scale, etc., are illuminated by small electric bulbs, the brilliance of which may be adjusted by a switch marked 'dimmer', 'scale lights', 'scale brightness' or 'illumination'.

Centring

As previously mentioned, when the electronic and mechanical centres do not coincide, errors will arise in the reading of bearings. Some types are provided with two controls, the horizontal and vertical 'centre spot' or 'shift' controls by which the whole picture may be moved in a horizontal or vertical direction.

Besides these there is usually another (non-electrical) method, namely changing the position of the focusing coil by mechanical adjustment.

Setting up the radar

The procedure for switching on and adjustment of the controls of each type of installation can be found in the manual supplied by the manufacturer. For most installations this sequence is:

1. Verify that the aerial can rotate freely, and that the sea-echo suppression and – with some British types – the brilliance and the amplification, are set to a minimum (turned in an anti-clockwise direction).
2. Switch to not too short a range.
3. Switch on the radar and wait about 2 min until transmission starts. (The switching of the relay after about 2 min is audible.)
4. Adjust brilliance in such a way that the timebase is just invisible.
5. Adjust focusing by observing a range ring halfway along the scale.
6. Adjust amplification until a faintly speckled background is visible.
7. If tuning must be done manually, adjust tuning control for exact tuning, using a meter or lamp for indication, and verify the exact tuning with the performance monitor (if present). If there is no indication for the exact tuning (with simple installations only), then tune on maximum extension of the sea-echoes.
8. Set the range switch to a short range and verify automatic frequency control, if present, by means of performance monitor.
9. Adjust sea-echo suppression so that there is a good contrast.
10. Set the range switch to the desired range and readjust focusing and gain.
11. Verify the centring of the picture and, if necessary and possible, adjust with shift controls.
12. Set the heading flash on $0°$ or on the ship's course (in accordance with the desired presentation).

When tuning to another range, the focusing, gain and sometimes also brilliance should be readjusted.

When going on watch, check range, gain, anti-clutter, tuning, centring, direction of the heading flash, and the entire performance, the last by means of the performance monitor.

PRESENTATIONS

So far it has been assumed that the heading flash always points upwards (0°). By controlling the orientation of the display with a gyrocompass, it may be kept so that the true north of the area represented remains in the same direction, irrespective of any changes of course by the ship. This is called azimuth stabilization. Moreover, if the display is controlled so that north of the centre of the display lies in the direction of 0° on the scale, it is said to be 'north-up stabilized' or 'relative north-up'. A presentation with the heading marker always at 0° is termed a 'ship's head-up' or 'relative course up' picture. When changing course with a ship's head-up

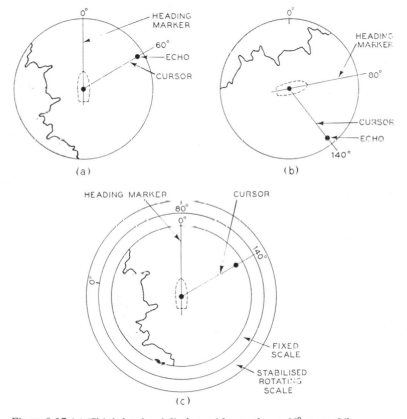

(a)   (b)

(c)

Figure 9.27 (a) 'Ship's head-up' display, with an echo at 60° green. When course is changed the display rotates but not the heading marker. (b) 'North-up stabilized' display, course 080°, echo at 140°. When the course is changed the heading marker rotates but not the display. (c) Stabilized second scale: the relative bearing can be read from the inner scale and the true bearing from the outer scale. The heading marker indicates true course on the outer scale; when course is changed, the display and the outer scale rotate but not the heading marker

picture, the picture turns but the heading marker remains in the same position, whereas with north-up stabilization the picture stays still and the heading marker turns. This is shown in *Figure 9.27(a)* and *(b)* in which the position of the ship with respect to the radar picture is indicated by a dotted line.

When navigating in narrow channels, where course changes often have to be made, it is dangerous to use the ship's head-up display. In such a display the echoes rotate gradually over an angle equal to each course alteration; because the previous echoes disappear slowly, the radar picture is unclear for some time. This is of course undesirable, particularly in narrow channels, and increases the danger of collision. The effect is called 'smearing' or 'blurring'. It also occurs when switching to another presentation or another range.

With ship's head-up, the yawing of the ship causes some blurring. With north-up the orientation of the picture remains the same when changing course or if the ship is yawing.

A third possibility is stabilization of the picture with the ship's intended course upwards (see page 299). If the ship is yawing the picture remains in the same position but the heading marker rotates.

According to the more experienced radar navigators, the north-up presentation should always be used. The navigator has to become thoroughly accustomed to it, but as soon as he is he will prefer to operate in this way.

Concluding, we may say that stabilization by means of the gyro is desirable for good radar interpretation. A great advantage is that more accurate bearings can be taken, because during course alterations the echoes remain in the same places on the screen. With ship's head-up, on the other hand, the position of the other echoes *is changed a little at each revolution of the scanner* when the ship is changing course or yawing. *This makes it more difficult to take accurate bearings.*

Changing from one presentation to the other is effected simply by reversing a switch with the two positions 'relative unstabilized' and 'relative stabilized' (ship's head-up and north-up, respectively).

True-motion radar displays

In the methods of presentation so far described, own ship always has a fixed place on the screen: the centre. In the true motion presentation this is no longer the case. It is now the fixed objects such as buoys, lightvessels, coastline, etc., that remain on the same spot on the screen. Our ship, on the contrary, is seen moving across the screen as other ships do.

The presentation is made by moving the centre of rotation of the timebase, which is the position of own ship in a stabilized north-up picture, across the screen. The direction of this movement of the electronic centre is the same as our course, and the velocity on the range chosen agrees with our speed. Clearly, geographically fixed objects will now no longer move across the screen.

The attractiveness of this presentation lies mainly in the fact that the movement of other ships on the screen is no longer relative to ours but is true. Hence it is called *true motion* or *true track* presentation. The name *chart plan* is also used. To explain true motion, the track of own ship is shown as AB in *Figure 9.28(a)*. At the times mentioned, own ship is in the position shown. CD is the path travelled by another ship.

By measuring the bearing and distance of this other ship we will first show the track of her echo when north-up presentation is used (*Figure 9.28(b)*). At time 1012, for instance, the true bearing is 85° and her distance 7 miles. The dotted line represents the track of the echo. As a result of the afterglow the echo will have a faint trace or short tail. *The tail indicates the direction in which the echo has moved* on the screen, i.e. the *relative* motion of the ship with respect to own ship. With a ship's head-up presentation, the motion of other ships with respect to ours is also relative. Hence both north-up and ship's head-up presentations are called *relative*. (The word relative here refers to the movement of echoes from other objects, not to bearings.)

In *Figure 9.28(b)* the closest point of approach (CPA) of the other ship is found by drawing a perpendicular, OP, from own ship (the display centre) to the projection of the other ship's relative course. The CPA can be found in the same simple way with a ship's head-up presentation.

When true motion is used (*Figure 9.28(c)*) we see own ship travelling the path AB and the other ship the path CD. They move along the direction of their courses. The 'tail' of both echoes shows their *true* direction. To exploit this advantage still more, the persistence (period of afterglow) of the screen is made longer, thus lengthening these tails, and the true courses of other ships are made still more visible. The greater the speed of the ships the longer their tails. *Figure 9.29* shows a photograph of a true-motion display.

The displacement of the electronic centre, and consequently of the whole picture, is obtained by mounting two extra pairs of coils around the neck of the cathode-ray tube, one pair vertically and the other horizontally. The currents flowing through these coils set up two mutually perpendicular magnetic fields. The electrons constituting the beam undergo two mutually perpendicular forces; these forces cause a displacement of the picture.

Thus the electronic centre no longer coincides with the mechanical centre (so-called off-centring). In accordance with the course and speed of own ship, both currents through the coils are gradually and automatically modified in such a way that own ship is always in the correct place on the screen. This necessitates an electrical coupling between the gyro compass

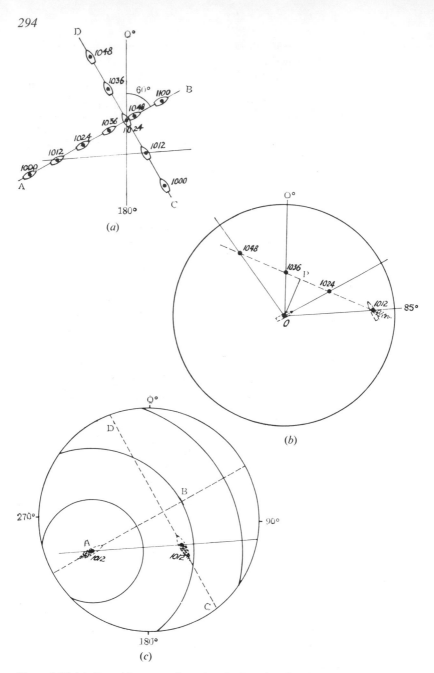

Figure 9.28 (a) Own ship moves from A to B, the other ship from C to D. (b) North-up presentation of (a); echo of other ship moves along broken line. (c) True motion presentation of (a): tails of own ship and other ship show their true, not relative, courses

and the radar, and also between the log and the radar. Gyro compass and log give information about course and speed to the radar installation in the form of currents and voltages. The change of current in unit time in the vertically and horizontally deflecting coils is made proportional to the required velocity components of the electronic centre in those directions.

When the distance from our ship to the edge of the screen becomes too small (for instance one-third of the diameter of the screen) and the important radar view forwards is reduced too much, the picture on the screen has to be reset from B to A in *Figure 9.28(c)*. The radar view backwards is less important, and for that reason the distance from A to the edge can be, say, one-sixth diameter.

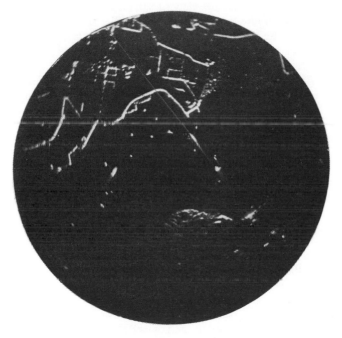

Figure 9.29 True-motion display: own ship (with heading flash) is in the channel marked by buoys and is on a course of about 330°. It is approaching the sharp bend near Bath (River Schelde)

The point A in *Figure 9.28(c)* should be chosen so that the heading flash passes through the centre of the screen. In that case it lasts for as long as possible before the picture has to be shifted again, unless the course is altered. When the range selected is, say, 10 nautical miles and the speed of own ship 15 knots, the track AB in *Figure 9.28(c)*, which is half the diameter or 10 nautical miles, will be travelled in 10/15 hour or 40 minutes. Every 40 minutes the picture must therefore be reset. To call attention to this a buzzer, and sometimes a lamp, are automatically brought into operation. If the picture were not reset, own ship would stop at the edge of the screen or sooner. The buzzer and lamp warn that the picture is then relative.

Two 'reset' controls are provided, one for shifting the picture vertically and one for shifting it horizontally, by altering the currents in the respective pairs of coils.

The resetting should not last too long as in the meantime echoes continue to come in and, owing to the prolonged afterglow, these echoes will remain some time. But even if the resetting is effected quickly the old picture will fade only slowly. Sea and rain echoes especially can be troublesome as they reduce the view of the radar in the forward direction. Some sets perform the resetting automatically. As soon as the current through one of the pairs of coils reaches a definite value, i.e. when own ship crosses an imaginary vertical or horizontal line on the screen, the picture jumps back.

By resetting, the navigator is deprived of good radar view for some time (say 20 seconds). The moment at which resetting has to take place may be inconvenient from a navigational point of view, so a facility has been incorporated into some sets of deferring this moment for some time. It is only necessary to throw a switch, in which case resetting takes place later on.

In a true motion presentation, the fixed and variable range rings (see *Figure 9.28(c)*) can still be used. However, in the area represented, the radar range in different directions from own ship is no longer the same as the range (picture radius) selected; the radar range is, according to the direction, longer or shorter and, moreover, changes gradually. It is better, therefore, not to use true motion when, for example, a landfall is made. In this case a relative picture is to be preferred.

Probably those conversant with relative plotting will choose a relative picture when meeting one or two ships in the offing, as this entails less work in plotting.

One can switch from true motion to a relative presentation, and back, as necessary. Both true motion and a relative presentation have their advantages and should be used according to the situation. An attractive but expensive solution is the installation of two displays, one giving a relative and the other a true motion picture.

It is clear that with true motion the mechanical cursor cannot be used by itself, because its axis of rotation is not above the position of own ship on the screen. We have to employ the electronic cursor and read the bearing from the special scale.

There is another method, however, enabling us to take bearings on a true motion picture without needing a special scale. On the perspex disc immediately in front of the screen a number of hairlines are engraved parallel to the central line that forms the mechanical cursor (*Figure 9.30*).

When the control of the electronic cursor is turned, this engraved disc is turned simultaneously so that the lines remain parallel to the electronic cursor. Therefore, when the electronic cursor is positioned so that it intersects an echo, the central graticule line of the mechanical cursor indicates the bearing on the *normal* scale.

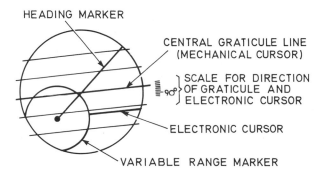

HEADING MARKER

CENTRAL GRATICULE LINE
(MECHANICAL CURSOR)

SCALE FOR DIRECTION
OF GRATICULE AND
ELECTRONIC CURSOR

ELECTRONIC CURSOR

VARIABLE RANGE MARKER

Figure 9.30 Electronic cursor and mechanical cursor (perspex disc engraved with parallel lines) are turned simultaneously by a single control; the central graticule line of the mechanical cursor, parallel to the electronic cursor, indicates the latter's true bearing. To prevent confusion the electronic cursor begins at the variable range marker

This method, moreover, facilitates the adjustment of the electronic cursor for, as we have seen, this adjustment is rendered somewhat difficult by the electronic cursor remaining on the same spot during one revolution of the scanner. The required rotation can now be judged from the position of the mechanical cursor, even though the latter does not pass through the echo whose bearing is to be taken. If the true course of another ship is to be determined, the parallel lines are turned so as to be as far as possible parallel to the tail of the echo. The central graticule line of the mechanical cursor then indicates on its scale the true course.

When the display is switched to a long range, the movement of ships on the screen is too slow to cause a sufficiently long tail. For this reason, the display is always made 'relative' when switched to a long range.

With most types, it is possible to switch from true motion to relative motion in such a way that own ship remains on the same spot of the picture but the movements of other ships over the screen alter from *true* to *relative* motion. With the latter, one can plot another ship's relative course (see later) and thus find its closest point of approach. The CPA can also be found (but less accurately) by forward extension of the tail after the echo has moved in the new direction for some time.

Operation of true motion

The various controls for true motion are normally mounted on the front panel of a special unit, a certain make of which is called the *Trackmaster*. This front panel is shown in *Figure 9.31*.

The desired presentation is selected by means of switch 1, 'Presentation', at the top right-hand corner. Assume that it is set to 'track indication' (i.e. true motion). A buzzer sounds when the picture requires resetting. The operator determines the point to which own ship is to be reset by means of switch 2; if this is set to, say, 'NE' the picture will jump, when switch 3 ('Press to reset') is pressed, in such a way that own ship is resituated at a point near the edge of the screen at 45°.

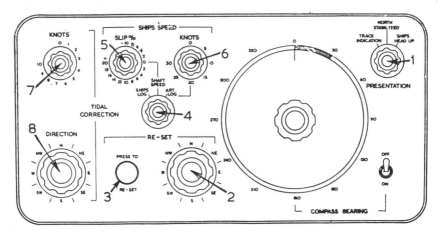

Figure 9.31 Marconi 'Trackmaster' control panel, showing operational controls

The position of the electronic cursor is indicated on the circular scale 'Compass bearing'. The toggle switch to the right of it should then be in the 'on' position.

By turning switch 4 to one of the following three positions the speed of own ship can be introduced in three ways:

1. By the *ship's log* in the form of electrical pulses. A fixed number of pulses per mile is supplied.

2. By a *tachometer* (or other instrument) indicating the number of revolutions of the propeller shaft per minute, and passing the information on in the form of a direct voltage or electrical pulses. Control 5 enables the slip of the shaft to be taken into account. The slip may vary without the number of revolutions per minute changing.

3. By a *pulse generator* (an 'artificial log') which imitates the log information. Control 6 permits the setting of the appropriate number of pulses generated per minute; it is set to the speed in knots at which the

ship is probably moving. This manual adjustment is provided in case speed information is not available from another source (e.g. owing to a fault) or for when true motion is to be adjusted or tested while the ship is lying in port.

It is evident that the fixed echoes will not remain at the same spot on the screen when the speed and/or course introduced are not correct. These echoes will leave small trails, all in the same direction, if the adjustment is inaccurate.

A better check is the following: suppose that on the six-mile range there are three range rings with two miles spacing. The outer ring may have a radius of, say, 150 mm. Six nautical miles range will therefore correspond to 150 mm on the screen. If own ship was travelling at 18 knots it would cover this in 20 minutes. The speed introduced should therefore be such that own ship travels 150 mm on the screen in 20 minutes. The direction of the track to be covered on the screen results from own course. Now we can mark on the reflection-plotter (see later) the point where own ship ought to be after 20 minutes. This is the best check.

Other presentations

In addition to the three presentations described so far (which are the most important) two others are occasionally encountered.

North-stabilized rotatable scale. Some radars have a second scale from 0° to 360°, mounted around the first; see *Figure 9.27(c)*. This scale is controlled by the gyro compass and is turned to, and kept in, such a position that 0° corresponds to the true north of the picture. The presentation is unstabilized (ship's head-up), and the gyro-stabilized outer scale rotates, together with the picture, in accordance with own ship's course changes. The heading flash, pointing upwards to 0° on the fixed inner scale, also indicates own ship's *true* course on the outer scale; using the cursor, relative and true bearings can be read simultaneously from the inner and outer scales, respectively.

Intended course upwards. The heading flash, together with the picture, can be turned to any position by means of a control marked 'Heading flash adjust', 'Picture alignment', or 'Rotary shift'. When the ship is heading precisely the course to be steered, the heading flash can be turned so that it points to 0°. The picture will then be stabilized in such a way that deviations from the intended course are indicated by a movement of the heading flash. The picture therefore shows no blurring (unlike the ship's head-up picture, in which the heading flash is steady and the picture turns). This presentation is called 'ship's intended course upwards, compass stabilized'; it is a *relative* presentation, i.e. movements of other ship's echoes on the screen are relative to that of own ship.

In one radar design, it is possible in the *true motion* presentation to turn the picture so that it is stabilized with ship's intended course upwards, instead of with north upwards. The heading flash will point to 0° when the ship is heading the intended course, and will move if the ship deviates from this course. This presentation is called 'true motion with ship's intended course upwards'.

The three positions of switch 1 in *Figure 9.31* are then as follows:
(*a*) Ship's head upwards, unstabilized.
(*b*) North upwards *or* ship's intended course upwards, compass stabilized.
(*c*) Track indication (i.e. true motion), with north upwards *or* ship's intended course upwards, compass stabilized.

In positions (*b*) and (*c*), the picture may be turned, by means of the alignment control, to the 'north upwards' or 'ship's intended course upwards' position, or if necessary to any other position.

RADAR SPECIFICATIONS

The performance of a radar depends on various interrelated specifications, which will now be discussed.

Pulse duration and pulse-recurrence frequency

In a previous example, it was assumed that a pulse with a duration of 0.1 μs was produced 1000 times per second (the p.r.f.). Therefore between the beginning and end of this pulse lies a time span of $1/10\,000\,000$ s. As the velocity of propagation is about 300 metres per microsecond, the start of the pulse will be about $0.1 \times 300 = 30$ m from the scanner at the very moment the end of the pulse is leaving the scanner. So when the pulse duration is 0.1 μs, a pulse about 30 m long is sent into space.

Range discrimination

Now suppose that this pulse meets an object of very small dimensions on its way. The echo is, of course, as long as the outgoing pulse (i.e. 30 m) and thus the length of the echo on the screen in a radial direction corresponds to 30 m. Note that the distance at which the object is located is immaterial.

Now assume that this pulse meets two objects on its way, P and Q, located in the same direction, one 15 m behind the other (the object P is so small that it causes no appreciable shadow in the outgoing radiation);

this is depicted in *Figure 9.32*. It will be clear that the beginning of Q's echo reaches P just as the end of the outgoing pulse is passing P. The echo pulses originating from P and Q are therefore contiguous, and the objects P and Q will not be made visible on the screen as two distinct echoes but as one 'merged' echo. If P and Q are more than 15 m apart, a separate echo will be seen from each. The size of the spot of light on the screen, which might affect unfavourably the observation of the two distinct echoes, is ignored here.

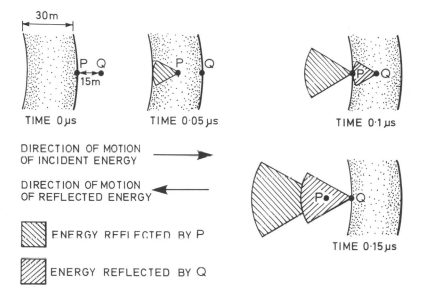

Figure 9.32 Pulse duration and range discrimination. Two targets, P and Q, are on the same bearing but differ in range by 15 m; the four diagrams show the progress of the incident and reflected energy of a 0.1 µs pulse (length 30 m) passing these two targets

The minimum distance in radial direction at which two objects will be reproduced separately is called the 'range resolution' or 'range discrimination'. It is equal to half the pulse length and is thus proportional to the pulse duration.

To be able to observe separate echoes from objects a short distance apart, the pulse should be short. It is now technically possible to generate reliable high-powered pulses of 0.05 µs and less. This corresponds to a theoretical range discrimination of 7.5 m or less. Most makes have a shorter pulse duration on the smaller ranges than on the longer ones (see under 'Pulse duration').

In practice the range discrimination of the various makes lies between 10 and 30 metres.

Pulse duration

On longer ranges, the discrimination is equal to more than half the length of the pulse, and is therefore less favourable. It is affected then by the size of the light spot on the screen.

The smallest possible diameter of the light spot is about 0.25 to 0.125 mm. If the radius of the spot is, say, 0.125 mm, the smallest distance on the screen between two echoes that can just be distinguished apart is 0.25 mm. With a 12 in ($\approx$ 300 mm) screen this corresponds to a distance of about 125 m on the longest (40 miles) range. On longer ranges there would therefore be little point in using a short pulse duration. A longer pulse means, moreover, increased energy of the outgoing pulse and of the echo, so that 'targets' hit by the beam will be better detected. In order to protect the magnetron against overloading due to a long pulse, the number of pulses per second is sometimes reduced automatically. On longer ranges the pulse duration is therefore in most sets longer and the pulse-recurrence frequency (p.r.f.) lower than on a smaller range.

There is one radar design in which it is possible by means of a special switch to choose between a pulse duration of 0.1 μs (p.r.f. 1000) and 1.0 μs (p.r.f. 500). This choice can be made only on the intermediate ranges.

When the aim is to make visible objects *at a large or a small distance*, a long pulse duration should be chosen so that the energy of the echo is increased. In the case of a short pulse duration, however, the lengths of the pulses in space (and thus the lengths of the echoes) decrease and a more detailed picture is obtained. This applies as well to faraway as to nearby objects. For detecting a buoy in sea-echoes, for instance, a short pulse duration is preferable.

Minimum range

It is evident that the pulse length also determines what is called the 'minimum range' of a radar set. It would not be possible to receive the complete echo of an object if the start of its echo had already returned before the end of the pulse train had left the scanner. For a pulse length of 0.2 μs, this minimum range is theoretically 30 m, but in practice it is longer. Moreover, the receiver is not able to detect an echo immediately after the very powerful transmitted pulse has left the scanner. The minimum range is consequently a little longer than half the length of the electromagnetic pulse; it may be assumed to be 15 to 30 m, depending on the pulse length and other factors. A short pulse duration is thus favourable to both range discrimination and minimum range. Technical development will probably shorten pulse durations still more in the future, but so long as pulses are used the minimum range will not be reduced to zero.

Range accuracy

The accuracy to which the distance from an object may be determined with the variable range marker is about 1½ per cent or better of the radius of the range to which the radar is switched. On the 12-mile range the error of any distance measured may consequently amount to 0.015 × 12 = 0.18 n. mile.

The range accuracy depends largely on the correct adjustment (see page 268) of the variable range marker with respect to the fixed range rings. The fixed range rings are more accurate than the variable one; their accuracy should be determined (for example when the ship is in port) by comparing the range of a radar-conspicuous point on the chart with the range as displayed on the radar screen.

Power

The power of radio-frequency oscillations generated in the magnetron during the transmission of the pulse is called maximum or peak power. If the same quantity of energy were generated continuously rather than in pulses, the power, of course, would be much less.

Horizontal and vertical beam width

A car's main headlights send out light in the form of a beam. Some of the light is also, however, radiated to the sides, as is readily noticed when standing alongside the car. It therefore becomes difficult to determine the width of the beam of light in terms of degrees. The same difficulty arises with the radar beam radiated by the directional aerial. *Figure 9.33* shows

Figure 9.33 Power in azimuthal direction. OB = OC = ½OA, B and C are half-power points; angle BOC is horizontal beam width, which is exaggerated here (it is usually only 0.6–2.0°)

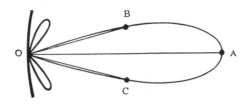

the field strength in azimuthal directions. In practice, it is agreed to take as the beam width the angle between the lines OB and OC, which are located so that the power output in these directions is half the maximum OA. Points B and C are called the 'half-power points'. (Instead of the angle between the half-power points a larger angle is sometimes given.)

The horizontal beam width of ship radar installations usually lies between 0.6° and 2°. When we look at *Figure 9.33* we should bear in mind that the beam width is much smaller than it is represented there.

If a small beam width is desired, the wavelength used must be small compared with the span of the reflector. This is of importance in the design of the installation. It is impossible to increase the dimensions of the reflector without limit, for this would cause difficulties of a mechanical nature.

In nearly all cases the transmitting aerial also serves for receiving. Now it can be shown that a diagram that shows the radiated power when the aerial is used for transmission also indicates the power received from various directions when it is used as a receiving aerial. It is evident that the directional effect of an aerial used for both transmitting and receiving is thus increased still more.

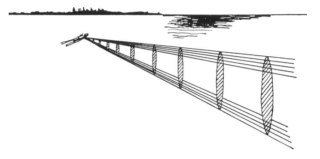

Figure 9.34 Ship with radar beam

The beam width in the vertical plane is made considerably larger. This is done to prevent echoes from being missed because of the ship's rolling and pitching. The angle for this vertical beam width is 15° to 30° (see *Figure 9.34*). Any desired beam width may be obtained by changing the shape and proportions of the scanner, just as any desired beam of light may be obtained by using appropriate reflectors. Radio waves are of similar nature to light waves and their reflection obeys the same laws. They differ only in wavelength.

Side lobes

Figure 9.33 shows the power in various directions in the horizontal plane and demonstrates that, in addition to the main beam, some energy is also transmitted in other directions. To eliminate these *side lobes* altogether, the reflector would need to be very large. In practice, a compromise is sought, and side lobes are tolerated to some extent. Instead of the two side lobes shown in *Figure 9.33*, there are usually a great many.

Obstacles located in the path of side lobes are also struck by the radar pulses, though these pulses have less energy, For clarity, the side lobes in *Figure 9.33* are shown too large compared with the main beam. The power in all side lobes together is usually about ¼ per cent of that in the main beam.

Later it will be shown that side lobes give rise to false echoes.

Lobes in the vertical plane

A closer examination of the radiated pattern in the vertical plane shows that it consists of a great number of lobes of the shape shown in *Figure 9.35*. The cause is the surface of the sea, which reflects the radiation.

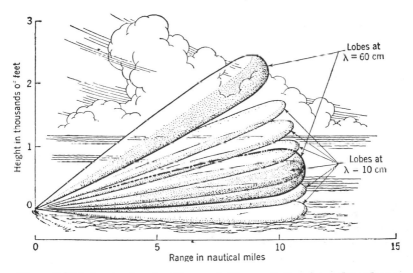

Figure 9.35 Lobe structure for 10 cm and 60 cm radiation over a flat reflecting surface (from Radar Aids to Navigation, *John S. Hall, courtesy McGraw Hill)*

Consequently the direct radiation ARP and the indirect radiation ABP will meet at a point P where they will wholly or partly reinforce or oppose each other (*Figure 9.36*). This depends on the difference in the path lengths ABP and ARP. What happens is comparable to fading due to the ionosphere, but note that a phase change of 180° takes place by reflection at B. If the difference in length between the paths at P were half a wavelength, corresponding to a phase change of 180°, then the total phase difference would be 180° − 180° = 0°. The two fields would thus be in phase at P.

Suppose that P' is located so that the difference in path lengths is one wavelength, then the phase difference will be $360° - 180° = 180°$. Here the two oscillations neutralize each other, at least if we take it for granted that after reflection at B' no loss is suffered so that the two fields at P' are equally strong.

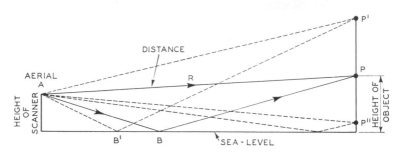

Figure 9.36 Direct and indirect radiation

For points immediately above the level of the sea, e.g. P'', the difference in path lengths is practically zero. Because of the phase reversal of about $180°$ on reflection, the two beams will neutralize each other at this point (even though they have travelled almost the same distance). *This is not favourable for the reproduction of very low objects.*

The number of lobes is in reality much greater than is shown in *Figure 9.35* and may amount to 500 for a wavelength of 3 cm.

When the sea is less calm (many sea-echoes) the reflection by the surface of the water is very diffuse and hence the minima between the lobes are, as it were, filled up.

To the navigation officer who, as a rule, is interested only in objects of small height such as ships, etc., the lowest of these lobes is the most important.

The lobes do not turn along with the ship when she is rolling or pitching. The difference in length between the paths at, say, point P in *Figure 9.36* is practically the same for a ship that is rolling and pitching as for one that is not.

Bearing discrimination

In *Figure 9.37(a)*, the horizontal beam width is such that the two equidistant ships A and B will be simultaneously struck by pulses, so that their echoes coincide. For the echoes to show separately the beam must be able to pass between the ships C and D without causing echoes (*Figure 9.37(b)*). The angle α between the ships must therefore have a minimum value somewhat larger than the horizontal beam width. This angle α is called the 'bearing discrimination' or 'bearing resolution', and is an important factor in a radar installation. It is clear that the angle in question does not depend on the distance at which the objects are located but chiefly on the beam width in the horizontal plane. In modern marine installations the bearing discrimination is from $0.6°$ to $2°$. Even in the best types of ship's radar,

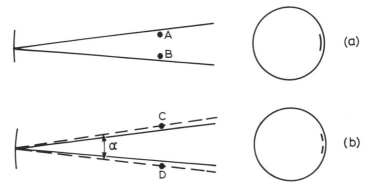

Figure 9.3 / Bearing discrimination. (a) Objects A and B are too close for the beam to pass between them, so they merge on the screen. (b) The angle between A and B is wide enough for the beam to pass between them without causing echoes, so two separate echoes appear on the screen

this discriminating ability is about 30 times worse than that of the unaided eye and about 300 times worse than can normally be achieved with the aid of optical instruments.

Distortion in azimuth

Suppose that an object such as a factory chimney is struck by pulses and that the horizontal beam width is $2°$ (see *Figure 9.38(a)*). When the reflector is in the position OA the chimney is already being hit, so an echo is perceptible on the screen in the direction OA. Succeeding pulses hit the object, and this is only discontinued beyond the position OB of the reflector. Hence the point-like object is made to appear on the screen as an arc subtending an angle equal to the beam width. If we want the arc to be small, i.e. reduce the distortion, a small beam width is necessary.

Now take the case shown in *Figure 9.38(b)*. The object in fact subtends an angle of $10°$, but on the screen its echo subtends an angle of $12°$ because the outer extremities of the object are already hit in positions OC and OD of the aerial. By reducing the gain, however, the object will be struck by little energy in the reflector positions OC and OD and the weak echoes will not be sufficiently amplified to be made visible on the screen; the echo-arc on the screen will thus become somewhat smaller.

There is another factor that influences enlargement of the subtended angle. An object with good reflecting characteristics such as a large ship needs to be struck by only a little radiation to reflect. A wooden mast, on the other hand, will give rise to an echo only when hit by powerful radar pulses. As a result the subtended angle is enlarged more for amply reflecting objects than it is for poorly reflecting ones.

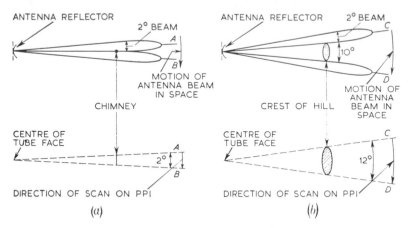

Figure 9.38 Distortion in azimuth: horizontal width of radar beam causes enlargement of echoes of (a) a chimney and (b) a long object

Since an echo's enlargement is expressed in degrees, its tangential dimensions will be greater the further it is from the centre of the picture. This can be observed on every radar picture. Consequently echoes from distant ships always give the impression that the ships are broadside-on (see, for instance, *Figures 9.21(b)* and *9.22(b)*). *Figure 9.39* shows variations in azimuthal distortion for ships at different ranges.

A small horizontal beam width offers the following advantages:
1. The bearing discrimination is improved.
2. The distortion in azimuth is lessened.
3. The energy will disperse less and remain more concentrated.

In order to obtain a sharp picture free of blurring and with all details clearly visible, the pulse length should be short and the horizontal beam width narrow. Pulse length and beam width therefore affect not only the two echoes behind each other (*Figure 9.32*) or side by side (*Figure 9.37*), however important this may be in some cases from a navigational point of view, but also the quality of the entire picture.

Bearing accuracy

A difference in alignment can arise between the motor shaft that drives the deflecting coils and the generator shaft driven by the scanner. It is clear that this can cause a bearing error which is slight in the case of accurate adjustment.

Bearing errors may also arise from the ship's rolling and pitching and the consequent movement of the beam. These errors are, however, so small that they need not be taken into account.

Thus the bearing accuracy of an accurately adjusted marine radar lies between 1° and 2°. In general terms, radar enables us to measure the

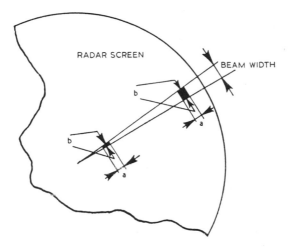

Figure 9.39 Variations in azimuthal distortion of echoes at different ranges: depths a are equal, widths b are not

distance to an object with reasonable accuracy, but a radar bearing is less accurate than a visual one. This is of importance in connection with radar position-finding.

Blind sector caused by the funnel

If the scanner does not protrude above the funnel, a blind sector must be taken into account. This is caused by the funnel intercepting the outgoing pulse and the echo, thus causing a radar 'shadow'. On some ships, this shadow can be observed on the screen, for instance when it cuts through a coastline. However, as the scanner is always mounted in front of the funnel, this blind sector will only impede 'radar view' behind, where it is usually not considered vitally important. The blind sector caused by the funnel may amount to an arc of from 10° to 45°.

Some shadow effect may also appear because of the foremast with its cross-tree and standing rigging. As energy is radiated over the entire width of the scanner, this shadowing is less than might be supposed, and is ordinarily present only on the short range scales. Nevertheless the intensities of the echoes of objects located dead ahead can be lessened by the effect of the foremast, as both the transmitted and reflected energy are weakened. This echo attenuation is such that it can reduce the maximum distance at which objects can be detected by a half or one-third. On normal cargo and passenger ships, with the usual distance between radar aerial and foremast, the sector of this weakening is small, from about 1° to 6°. This does not mean, however, that this blind sector can simply be ignored.

By mounting the scanner in front of the foremast or between the masts on the cross-tree of a twin-mast, shadows ahead can be eliminated.

When the ship is in port it is impossible even for experts to determine with certainty the precise effect and size of the blind sector or the sector of diminished view. It is therefore highly recommended that the master checks this at sea.

This can be done by observing to what extent the echo of a poorly reflecting object, for instance a small buoy, is affected when it goes into the blind sector of the display. Sea-return can also be used. A strongly reflecting object, for instance a coastline, is not suitable because its echo-energy, even when in the blind sector, is usually still sufficient after amplification to be reproduced on the screen.

If there is shadow-effect in the direction ahead, it is possible when navigating in fog to obliterate the blind sector by making small changes in course to port and starboard. The extent of the change in course depends on the size of the blind sector.

Maximum range

It often happens with a high sea that when echoes of distant ships are observed an echo will not appear on the screen with every revolution of the scanner, but can be missing for one or more revolutions. Because of the ship's pitching and rolling and the mass of water between the two ships, the pulses may not (or not adequately) strike the other ship for several scanner revolutions. The same holds true, at a smaller distance, for buoys. This is no indication that the installation is operating unsatisfactorily. This phenomenon is also caused by the fact that the reflecting power of some objects is subject to directional effects. Take, for example, a metal plate. Clearly this will reflect to a maximum extent if the radiation by which it is struck is directed perpendicular to the plane of the plate; the reflection will be reduced to almost nothing if the plane is parallel to the direction of the radiation.

The same effect can occur with ships, and even more often with buoys. Depending on the angle of incidence, the reflecting power will be greater or smaller; in the lapse of time between two consecutive 'irradiations' (2 to 10 seconds), differences may have occurred in the angle of incidence and reflection.

Later it will be seen that the distance at which an object is first observed is, in some measure, a means of ascertaining what kind of object it is; for instance, a buoy or a ship. This distance is usually taken as that at which the echo appears on the screen for 5 out of 10 scanner revolutions.

Number of scanner revolutions per minute

In every revolution, radar pulses from the rotating scanner must strike the object not once but repeatedly during the time it is aimed at the target. The effect on the screen of the repeated strikes will be cumulative, i.e. every echo-pulse received enhances the light-spot on the screen. Should insufficient pulses hit the target only a weak light-spot will be produced and this will disappear too quickly. The required number of pulses depends largely on the afterglow of the c.r.t. screen material.

The number of revolutions per minute of the scanner is between 20 and 30.

Wavelengths used in radar

If the transmitted electromagnetic energy is to be sufficiently concentrated in a beam without employing large reflectors, it is necessary (as already stated) to use very short wavelengths. Most marine radars make use of the X-band (wavelengths of 2.7—5.8 cm, i.e. about 3 cm).

The S-band (about 10 cm) is less frequently used. With this wavelength it is more difficult to obtain a narrow beam, as a result of which the picture is usually less detailed. On the other hand, radiation on a 10 cm wavelength passes better through rain showers and causes less sea-echo. For this reason some ships are fitted with a 10 cm as well as a 3 cm radar.

There is a third radar band, the Q-band, with about 8 mm wavelength. Unfortunately this wavelength is not suitable for obtaining a sufficient range owing to excessive attenuation in the atmosphere; the range is limited to about 4 nautical miles.

Linear and circular polarized radio waves

In some radars it is possible to transmit either circular- or linear-polarized radio waves. Reflection depends on the polarization direction; circular-polarized waves give less precipitation-clutter and sometimes also less sea-clutter on the display.

EFFECT OF ATMOSPHERIC CONDITIONS ON RADIATION

It is well known that the rays of the sun are somewhat refracted by the atmosphere towards the Earth. The radiation from radar transmitters undergoes a similar refraction. It is due to the gradual change in refractive index with increasing height, caused by variations in temperature, pressure and humidity of the atmosphere. The refractive index referred to is the ratio of the velocities of the rays in two successive media, the media being layers of air at different temperature, pressure and humidity. The refraction of radar signals under normal conditions is greater than that of light, owing to the much longer wavelengths of the radar waves. This has the effect that *the range of a radar set is greater than the optical range*. The so-called *radar horizon* is located about 6 per cent beyond the optical horizon and about 15 per cent beyond the geometrical horizon. For ships with an aerial height of 20 metres the radar horizon amounts to about 10 nautical miles.

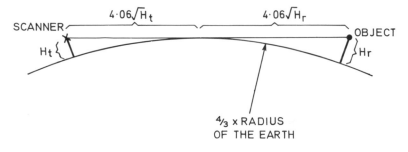

Figure 9.40 Average radar range

    Instead of basing calculations on a curved radio-wave path, it is easier to suppose that the radius of the Earth is 4/3 times its actual value; in this case the propagation of 3.2 cm waves may be assumed to follow a straight line. The distance to the radar horizon will then be found more easily (see *Figure 9.40*).

The approximate range may be computed from the formula

$$D = 4.06\sqrt{H_t} + 4.06\sqrt{H_r}$$

where D is the range in kilometres, H_t is the height of the scanner in metres and H_r is the height of the object in metres. It is evident from *Figure 9.40* why the formula is symmetrical in H_t and H_r. The formula has, however, only a limited significance, as it applies only at one specific temperature and pressure at sea level, and assumes a certain decrease of both with height and a certain unchanging relative humidity. The formula is in fact based on the averages of the quantities mentioned for the whole world. Furthermore this formula does not take into account the lobes due to reflection by the sea-surface.

Sub-refraction

Under special atmospheric conditions the refractive index can change so that the radiation is either bent downwards less or even bent upwards. If a diagram is drawn as in *Figure 9.40* with the radius of the Earth assumed to be 4/3 times its actual value, the rays may be bent upwards as shown in an exaggerated way in *Figure 9.41*. This is called *sub-refraction*. It will be

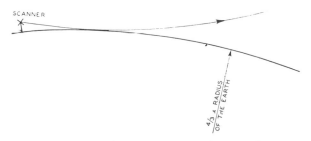

Figure 9.41 Sub-refraction produces shorter radar range

seen that the range is now smaller. Sub-refraction may occur at high latitudes, when a cold mass of air flows over a warmer surface (for example, wind blowing over open water after having passed over ice).

Super-refraction

In *Figure 9.42* the rays are bent downwards more than in a so-called standard atmosphere; this is called *super-refraction*. Super-refraction may

Figure 9.42 Super-refraction produces longer radar range

occur if the air, after passing over a warm land mass, flows over a relatively cold sea. The greater the temperature difference between air and sea the more this phenomenon will appear. It can thus be expected in the vicinity of land and in temperate and tropical zones. Yet it can also be present on the open sea, far from the coast, especially in the regions of the trade-winds. The effect of super-refraction is that the range may become greater than is normally the case, as is shown by the comparison of *Figures 9.40* and *9.42*.

Super-refraction is encountered quite frequently. In the English Channel, for instance, the wind usually comes from over the land. During the spring and summer, when the land temperature is usually higher than that of the water, super-refraction can be expected during the greater part of every 24-hours. In the Mediterranean, also almost completely surrounded by land, super-refraction has been observed nine days out of ten during the spring and summer.

If the rays are bent down still more, they may be reflected by the surface of the land or the sea, then curved down again, reflected again, and so on (see *Figure 9.43*). This process is called *ducting*, the rays seeming to

Figure 9.43 Ducting of radar waves

be confined in a duct. Extremely long ranges of hundreds of nautical miles may occasionally be obtained.

Ducting can also occur at heights of several hundred metres: two adjacent layers of air may have such different refractive indices that the lower layer repeatedly reflects radiation rather than refracting it. In this way a radar transmission or its echo can be propagated a very long distance (sometimes hundreds of nautical miles) before escaping downwards. Apparently inexplicable radar pictures can be brought about, albeit very seldom, by this phenomenon.

Echoes of precipitation and clouds

The water, whether or not frozen, present in the atmosphere as fog, clouds, snow, hail or rain absorbs energy to a greater or lesser degree. This means that some of the transmitter energy and the echo is attenuated, and that objects behind or within an area of mist, clouds or precipitation will produce weaker echoes. Also clouds and precipitation themselves can give echoes that are sufficiently strong to be visible on the screen (*Figure 9.25(a)* and *(c)*). With fog, snow and very light rain this is not the case. If the intensity of the precipitation is sufficient, the echoes on the screen will have the appearance of speckled blots with vague edges resembling wool, and when the echoes are still stronger there is a uniform illumination with well-defined edges. If the edges are vague this helps to distinguish them from land echoes. Sandstorms, which can limit normal visibility drastically, have little effect on radar view. Very heavy sand-storms (wind-force 9), however, show faint uniform echoes which may mask echoes from other ships, etc.

DANGERS OF RADIATION

Radiation from aerial or waveguide

The danger from radiation at very short wavelengths has been investigated very thoroughly by at least three independent groups of experts. After a lot of trials on animals (e.g. mice in waveguides) and measurements of radiation the conclusion may be concisely summarized as follows.

Electromagnetic radiation from the aerial could only be dangerous for a person standing for a long time in the main beam of a non-rotating aerial and at a distance of one metre or less; there the average power can be 10 mW/cm^2. Radiation direct from a waveguide is even more concentrated; when making repairs, for instance, one should avoid standing in front of (and especially looking into) an open waveguide.

At 3 cm wavelength, a sense of heat on the skin surface warns of the presence of radiation. At 10 cm wavelength, a rise in temperature occurs at about 2 cm below the surface of the skin; this is not noticed, and radiation at this wavelength could be dangerous for the eye-lens amongst other things.

Röntgen radiation

It is well-known that when fast-moving electrons hit the screen of a cathode-ray tube (radar or television), Röntgen radiation is produced. The intensity of this radiation is very limited, however. During close investigation both in the United States and in the Netherlands (Royal Navy), no case of permanent injury from this radiation has been found.

Radioactive radiation

Some components (trigatron and T-R switch) exhibit slight radioactive radiation. At a distance of over one foot this radiation is negligible, but one should be careful not to carry parts of these tubes in one's pockets or clothing.

RADAR AS AN AID TO NAVIGATION

To be effective in preventing collisions, it is of the utmost importance that the radar screen should be observed regularly and methodically. It should be borne in mind that new echoes, sometimes at close range, may appear at any time. The number of echoes, their more or less dangerous approach, the quantity of sea and/or precipitation clutter, and other conditions, should all be taken into account when deciding at what intervals the screen should be observed or whether continuous observation is necessary. If there are indications that visibility will suddenly deteriorate the radar installation should be switched on or put to stand-by. In regions where fog banks, small craft or perils such as icebergs may be expected, the radar should be continuously switched on. This is most important in the case of fog banks, because other ships can then be identified and radar tracking commenced, before they disappear into the fog banks.

It is important that the captain and others using the radar acquire experience in observation and interpretation of the picture and in navigation. Moreover they should keep this up by practising in bright weather conditions. Under these circumstances radar observations may be compared with visual observations and a wrong or an erroneous judgement of the situation need not incur danger.

As to the choice of presentation and the positioning of the display(s), we may refer to an article by Captain A. Wepster (Journal of the Institute of Navigation, July 1963). The solution given in this article is the use of two display units, one of which has a relative north-up presentation and the other a true-motion one. A range that has proved to be favourable in practice for the relative display is 12 nautical miles and for the true-motion picture 6 nautical miles. The more important of the two is the relative picture. By this method one can pick up among other things ships at a great distance, follow their path and determine their closest point of approach.

The true-motion picture, in which we see the true course of other ships and also obtain an impression of their speed, gives an insight into the navigation situation and facilitates taking a decision about a manoeuvre. It is desirable that the diameters of the two screens are at least 12 inches ($\approx$ 300 mm).

The displays are best located in a separate room (plot room) that is not quite dark but illuminated to the faint level of the radar screen. The observer should be able to look out to the front, e.g. by pushing aside curtains before a window. In the room there should be an illuminated clock, a remote control for v.h.f. telephony, an indication of the speed of the ship and a table for making notes and for plotting.

If no separate plot room is available, the screen of the radar indicator unit can be provided with a hood or visor to prevent daylight from falling directly on the c.r.t. screen. Undesirable reflections are thus avoided and the contrast is improved. Disadvantages are that only one person at a time can look at the picture and that plotting on the reflection plotter (see page 333) is rendered more difficult (if not impossible). An aperture is sometimes provided in the hood and plotting may be achieved by inserting the hand holding the pencil through this.

Influence of various materials on the strength of the echo

Many targets from which echoes might be expected will not show up on the screen and it is important to know the cause.

Rays of light that hit an object are reflected, absorbed or (as with glass) transmitted through it. The same holds for radar waves; there is always some absorption. Few materials readily pass radar waves; one example is the special cover over the waveguide extremity or slotted waveguide of the aerial.

It is, of course, the reflection that is the most important. This depends to some extent on the material of the reflecting object. The reflecting properties of metal and water, for instance, are better than those of wood, stone, earth and vegetation such as trees, etc.

Influence of shape and aspect of the target on echo energy

A beam of light is reflected in a single direction ('specular reflection') only when the reflecting object presents a very smooth surface, e.g. an unruffled sheet of water or polished metal. The dimensions of any surface unevenness must be small compared with the wavelength. The wavelength

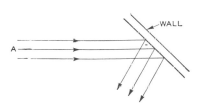

Figure 9.44 Radiation from transmitter A hitting a wall at such an angle that there will be no echo

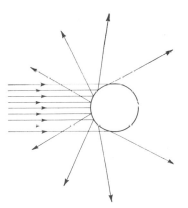

Figure 9.45 Diffused reflection by sphere

of the average shipborne radar set is about 50 000 times longer than that of light; thus a surface such as a wall that would scatter light in all directions (diffused reflection) behaves like a mirror to radar waves. If the radar waves strike such a surface at an angle, the reflected rays will not be returned to the receiving aerial (*Figure 9.44*). If the radar waves are perpendicular to the surface, however, the echo is very strong.

In contrast with the 'specular' reflection in *Figure 9.44*, the diffused reflection of a sphere scatters the radiation uniformly in all directions (*Figure 9.45*). Between these two extremes there are all sorts of inter-mediate conditions in which reflection takes place chiefly in one direction but with still sufficient reflection in the other directions (*Figure 9.46*). The more rays are reflected back towards the receiving aerial, of course, the better the reproduction of the target.

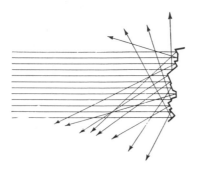

Figure 9.46 Irregular surface causing reflection in one main direction

Consider three planes at right angles to one another (*Figure 9.47*). It can be proved that in this case the direction of the reflected radiation is always *exactly* opposite to that of the incident radiation. The trihedral angles (corner reflectors) mounted on buoys and other navigational targets make use of this property.

The reflecting power of a trihedral angle is so strong that, even if the dimensions of the planes are small, sufficient echo energy will be produced. Dihedral and trihedral angles can also be formed by, say, two or more

Figure 9.47 Three metal plates at right-angles to one another always reflect rays in the direction from which they came

houses with or without the surrounding ground, or by rock formations, which then give an excellent echo. This is why the echoes from blocks of houses are usually of such good quality for radiation from all directions (*Figure 9.48, 9.49* and *9.50*).

In contrast to the strong reflecting power of trihedral angles, the com-pletely smooth sloping surface of an iceberg may reflect radiation chiefly

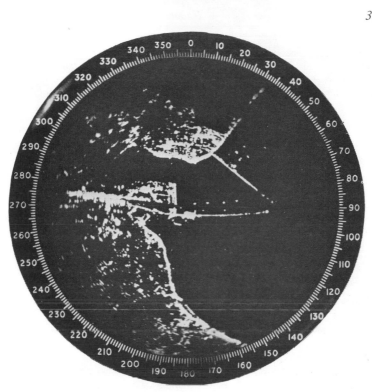

Figure 9.48 Approach to the mouth of the R. Liffey and Dublin, showing the breakwaters and channel buoys, on the screen of a marine radar set to the 3-mile range (courtesy Marconi International Marine)

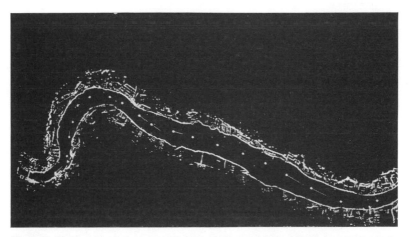

Figure 9.49 Mosaic of displays: Liimfjord, Denmark (courtesy Decca Radar Ltd)

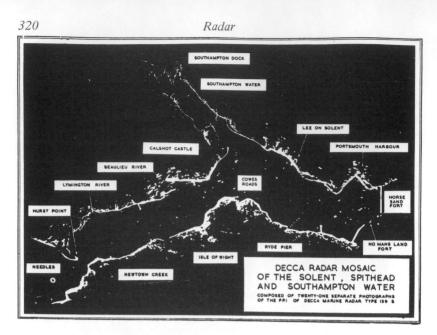

Figure 9.50 Mosaic of displays: Solent, Spithead and Southampton Water (courtesy Decca Radar Ltd)

upwards and the iceberg may not be detected. However, there are usually some reflecting points even on the apparently smooth and flat surfaces of a dune, a hill, a cliff, a mountain and even an iceberg. With a few exceptions, such objects still therefore appear on the screen provided that the transmitter is not at too great an angle to the perpendicular of the reflecting surface.

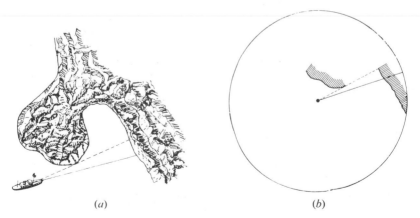

(a)   (b)

Figure 9.51 The bay shown in (a) does not appear on the screen (b)

As the ship moves, the angle of incidence of its radar radiation will change continually, and new points will cause reflection. Thus the screen image of a coastline, and the land beyond, is constantly changing.

It should further be borne in mind that hills, etc., may screen flat country or lower hills behind. Thus the bay of *Figure 9.51(a)* will not appear on the screen (*Figure 9.51(b)*). Similarly, a mountainous island (*Figure 9.52*) will, as it were, cast a shadow on the lower country beyond. Even a big ship at close range may cause a shadow. In *Figure 9.53*, for instance, the *Queen Mary*, approaching Southampton, intercepts the radiation from a shore-based radar and throws a shadow on the shore (in the direction of about 80°).

Figure 9.52 Recognizing a coastline on the screen: island prevents reception of echoes from the land behind

SHIP

The shape of a lighthouse is often conical, and hence it reflects radiation in a slightly upward direction, even if the vertical angle of the cone is small. Such a target will therefore show up less well on the display than would be desirable from a navigational point of view. This applies to an even greater extent to some conical buoys.

Sandbanks, faintly sloping beaches and other low land with a smooth surface reflect the radar radiation more or less in a specular way. Because of this and their small height they will not always appear on the screen, especially not at long distances. The less smooth their surfaces are, the more the reflection will be diffused and the better they will appear on the screen, and at distances of a few miles or less they generally produce excellent echoes.

Influence of target size on echo energy

As yet, we have only considered the influence of the shape and the aspect of an object on the echo energy. Obviously the size of the object affects the echo energy, and a large ship produces a stronger echo than a small one at the same distance. Use can be made of this. If, by reducing the gain, an echo on the screen is weakened more than another echo at the same

Figure 9.53 Display showing the Queen Mary *under way to Southampton; NW. Netley buoy to starboard, a number of small vessels ahead. At 80° the ship casts a shadow over the shore. Picture obtained by a shore-based station, not situated in the centre of the screen (courtesy Decca Radar Ltd)*

distance, the first echo is a smaller target than the second. In this way it is sometimes possible to isolate some object on land the echo of which is merged with that of the surrounding country, or to distinguish the echo of a buoy from that of a ship in its immediate neighbourhood.

Switching

When the radar is switched to a long range, it should not be left at this setting when there are other ships at close range, as any manoeuvres of these ships would be noticed too late. The range to which the radar is switched should normally be approximately equal to the ship's speed.

When switching over, the old picture will slowly fade out, especially if the tube has a long afterglow. If necessary, the new picture should be given a little more brilliance enabling it to be distinguished from the old one more quickly.

It is also possible, before switching to another range, to reduce the gain so that the picture slowly fades. (During this time the picture can be turned up again if necessary.) After the picture has disappeared we should switch over. In this way little or no blurring will occur.

A good method of guarding against ships approaching dangerously (e.g. to 3 miles) is to set the variable-range ring to 3 miles to act as a 'danger warning circle'.

Radar detection of icebergs and ice fields

The extent to which the radar installation can indicate the presence of ice fields and icebergs in the environment is of importance. The following data taken from various reports may give an idea.

According to observations made by ships plying between Greenland and Newfoundland, the maximum distances for detection (depending on weather conditions and the dip of the horizon) are: 48 to 14 n. miles for very large icebergs, 40 to 10 n. miles for medium-sized icebergs, 36 to 7 n. miles for small icebergs and 15 to 4 n. miles for 'bergy bits'.

Just as the surface of the water does not produce echoes in calm weather, neither does the smooth surface of an ice field. But when the ice surface is not smooth echoes will be produced in proportion to its degree of roughness. If the movement of the sea causes the ice field to disintegrate into pieces that later freeze together again, a large number of rough blocks of ice are formed, each block bordered by practically perpendicular surfaces that act almost like corner reflectors as far as the production of echoes is concerned. Such a rough ice surface almost always produces very clear echoes to a distance of 2 or 3 nautical miles.

Smooth ice fields may be distinguished from open sea, because the sea produces sea-return with full gain while the ice fields do not. However, lumps of pack-ice and other irregularities in an ice field are perceptible. Echoes recurring in the same spot then indicate ice formations, whereas the ever-changing position of sea-echoes denotes open water.

'Growlers' (very small icebergs) are only perceptible on the screen at a few miles distance; when there is appreciable sea-return they may pass unnoticed. They protrude only about 2–3 m above water, but on account of their large submerged mass they constitute a hazard that will not in all circumstances be indicated on a radar display.

Ice can be a problem in radar position-fixing (see later): the coastline appearing on the screen may be altered by any icebergs or pack-ice located close to the coast or stranded on it.

Maximum distance

The maximum distance at which objects are still visible on the screen can give an idea (if only a very rough one) of the size and nature of the objects, and thus give clues for identification. Hence a representative list of maximum distances for an aerial height of 15 m is given below. This information can be of use, among other things, to determine one's position when making a landfall.

Hills or cliffs higher than about 30 m	15 nautical miles or more
Ships (about 12,000 tons)	12–17 nautical miles
Small ships (about 1,000 tons)	6–10 nautical miles
Lightvessels	6–10 nautical miles
Fishing craft	3–9 nautical miles
Very small buoys	½–1 nautical mile
Very large buoys	4–6 nautical miles
Radar reflectors on large buoys	6–8 nautical miles
Low coast, a few feet above water	1–6 nautical miles

These distances are maximum distances in the absence of super-refraction and sub-refraction. During fog they can be 15–20 per cent less.

Radar position fixing

Position fixing with radar is based on the principle that, if the distance to a fixed point is known, the locus of the ship's position is a circle with a radius equal to the distance and with the fixed point as centre.

As previously mentioned it is difficult to identify land echoes with certainty at a great distance. If this is indeed the case, the procedure for position fixing is that, to begin with, the bearings and distances of three or more echoes are measured; these three distances and bearings are then plotted from a given point on transparent paper. The distances are drawn to the scale of the Mercator chart for the relative latitude. The paper can now be placed on the chart so that true north on the chart coincides with true north on the paper and the points from which the echoes come coincide with points on the chart from which such echoes could be expected. For this operation the 'Radar Station Pointer' (Admiralty Chart 5028) can be used with advantage.

The previously described distortion in azimuth of echoes on the screen, especially of those located near the edge of the screen, also occurs with coastlines. In *Figure 9.54* a true coastline is shown with the radar representation (broken) superimposed. On the screen the approach to the bay D appears narrower than in reality. Cape A is made visible as an arc BC. By taking bearings and range from the centre of this arc with the aid of the cursor and the variable range ring, such a point can be used for position fixing.

There will be no distortion only if the coastline on the screen is at right angles to the cursor, but this does not necessarily mean that such a coast is always suitable for taking a bearing (e.g. point F). If E is the only point from which a bearing can be taken, allowance should be made for the fact that distortion has caused this bearing to shift by an angle equal to about half the horizontal beamwidth. But all distortions are lessened as the gain is reduced (page 307).

If there are three conspicuous radar landmarks, the position can be found by measuring the three distances, and if the three distance circles intersect on the chart or form a triangle (provided that it is not too large),

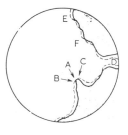

Figure 9.54 Distortion in azimuth of a coastline (exaggerated). Solid line is true coastline, broken line is the coastline seen on the screen

a check with the bearings is not necessary. It is, however, very important that the bearings of the three landmarks should differ greatly from one another. The bearing and distance of a small isolated object such as a light-vessel or buoy can be measured more accurately than those of, say, a cape. A difficulty with the bearing of a cape is to tell whether the cursor runs exactly through the centre of the cape (points A and F in *Figure 9.54*). With the distance it is not always known which point on the chart corresponds to the nearest point of the land that appears on the screen.

PLOTTING

Radar gives us a faithful picture of our surroundings. There are, however, some characteristic differences from visual observation that are of importance with regard to plotting.

Firstly the radar picture is not seen from the centre, as with visual observation, but from a short distance in front of the screen. Consequently changes in bearing on the screen will be observed by the eye less quickly than they would by visual observation. This is shown in *Figure 9.55* where angle β is smaller than angle α. This drawback disappears, however, when measurements are made with the aid of the fixed or the electronic cursor from the centre of the screen.

When a change of distance is to be observed, radar clearly has the advantage over visual observation. Such a change is detected best by the aid of the fixed or variable range rings. Another method, marking the positions of the echoes on paper or a reflection plotter, may also be applied. Tracing the positions of an echo is called plotting. If plotting is effected at regular intervals, not only changes in bearing but also in speed may be observed in a simple way.

The relative radar display gives us only the position of another ship and not the aspect. To determine aspect, which is often essential to form a judgement of the situation, plotting is required.

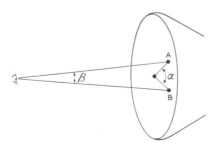

Figure 9.55 Movement of another ship from A to B (change in bearing α) is seen by the eye as movement through the smaller angle β

Finally, it should be borne in mind that every radar is subject to inaccuracies. The accuracy in bearing is usually $1°-2°$ and the error in range can amount to a small percentage. Then there are the errors in observation. Thus in practice an absolutely accurate display of the situation can never be expected, but we can obtain a picture sufficiently accurate for working purposes. To keep errors within reasonable limits, observations should always be made with the maximum possible accuracy.

Plotting can be done on the reflection plotter or on paper (manoeuvring board); see later. The purpose of plotting is the determination of a possible collision danger and of an avoiding manoeuvre, if necessary.

True plot

Various plotting methods will now be discussed. The first is the *true plot*. A line can be drawn representing own course and speed and, from the successive positions of own ship, the bearings and distances of another ship. A plot as shown in *Figure 9.56* is then obtained. This plot gives a clear picture of the position of both ships and also shows directly the course and speed of the other ship.

There are, however, some shortcomings to this system. Firstly, we cannot see immediately whether there is danger of collision. This can be determined, it is true, by checking the bearing, but this entails an extra operation. Secondly, it is difficult to say what will be the closest point of approach and when the ship will be at this range. These difficulties are even more pronounced when the effect of any avoiding action has to be determined.

For a trained observer, relative plotting offers more possibilities than true plotting. For an observer lacking experience in interpreting relative

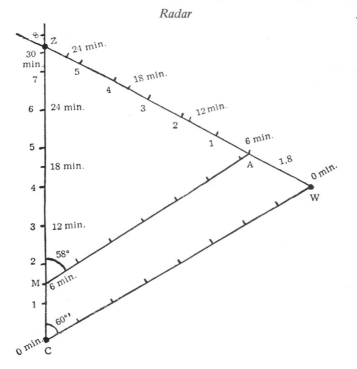

Figure 9.56 True plot of movement of own ship (C to Z) and another ship (W to Z)

motion, however, the true plot is preferable. One should never forget that if time for a true plot is lacking, all the information needed is supplied by the relative picture.

It is intended to obtain from the radar certain information; the method used depends on the number of people on the bridge, the radar installation and also on the personal preference of the Master. Thus the choice of the plotting method is not so important, provided that it leads to an exact conclusion.

Relative plotting

In a relative picture, own ship remains at a fixed point on the screen. We cannot see our own movement. What we see, just as when observing the surroundings visually from the bridge, is the relative movement of the coast, buoys, other ships, etc. On the screen, the echo of a stationary object moves at own ship's speed, on a course opposite to own course. The echo of another ship has own ship's course and speed 'subtracted' from its course and speed; the relative movement is the vector-difference of both movements.

Own ship remains, therefore, in the same place and all echoes of other ships and land move relative to us. Thus if the movements of ships' echoes are plotted as they appear on the picture, a relative plot is obtained. This is usually done on a plotting sheet, which is a diagram with range rings and radial lines to the graduated outer circle (*Figure 9.60*).

It should be noted that relative plotting is less accurate with a non-stabilized (ship's head-up) picture than with a stabilized (north-up) picture, for the following reason. In the non-stabilized picture, when own ship changes course or yaws the picture follows, and the heading marker continues to point to $0°$. The bearings of all echoes are measured relative to own ship's head, and must be converted to true bearings (by adding own ship's course) in order to be plotted on the plotting sheet. It is doubtful, however, whether own ship was steering the exact course at the moment of observation. The echo is repainted once every 3 or 4 seconds. A yawing ship may turn as much as $1°$ per second, and it would be mere chance if own ship was on the correct course at the moment of observation.

Example

In *Figure 9.57* the speed of own ship is 12 knots and the course is $30°$ true. At 0800 a ship is observed at point O, and at 0810 the ship is at point A. If we draw a line through O and A this is the line of relative approach, indicated by a ringed arrow. The distance OA is travelled in 10 minutes, so we know the relative speed of approach and can determine the moment at which the ship will have reached the closest point of approach, point N. If OA is equal to 2 and AN to 7 nautical miles, the ship will be at point N at 0845. If the other ship had no movement of its own it would have moved along the line OW parallel to our own course and after 10 minutes it would have been at point W; OW is 2 nautical miles. The other ship, however, is moving, for it is not at W but at A. The other ship's own motion during these last 10 minutes has consequently been the line WA. These three lines now form the plotting triangle. From W two true courses, own course and speed WO, and the other ship's course and speed WA, are drawn. We mark them with an arrow. The difference between these two vectors, the line OA, is the relative motion.

We now know that the other ship will pass ahead of us at the closest point of approach of 1 nautical mile at 0845. Course and speed can, if necessary, be measured in the triangle, but usually this is not done. By comparing the line WA with our own speed WO we already get a good impression of the other ship's speed.

What we want to know in addition is her 'aspect', i.e. the bearing of own ship taken from the other ship (angle BAM in *Figure 9.57*). This angle is measured from dead ahead to dead astern with the addition of red or green for port or starboard. It is found by taking the difference between the other ship's opposite bearing and her course. In *Figure 9.57*, therefore, the aspect is 70° red.

Together with the bearing, the aspect at once gives us a clear picture of the situation. In visual navigation we assimilate these facts subconsciously.

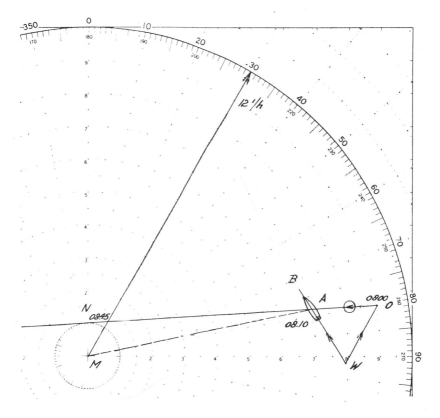

Figure 9.57 Relative plot on a north-up presentation. WO = Way of Own ship, WA = Way of Another ship

In radar navigation in fog, one man observes the screen, so the observations have to be passed orally. Bearing, distance and aspect are very well suited to this purpose.

The relative plot described here has various advantages. It takes little room and its construction is simple. It supplies the data mentioned above and, lastly, it enables us in a simple way to construct new lines of approach when we are preparing to manoeuvre.

Speed alteration

Consider a speed manoeuvre. In *Figure 9.58* own ship's course is 10°
true and her speed 15 knots. At 1000 ship B was observed at point O and
at 1008 at point A. During these eight minutes own ship has travelled 2
miles, so we can find the point W. The line WA consequently shows the
course and speed of the other ship during the plotting period. It is decided
to let ship B pass 3 miles ahead by reducing our speed. At 1012 we begin
reducing speed and we estimate that by 1020 we can get down to the new
speed. We now have to determine 'what should this speed be?'

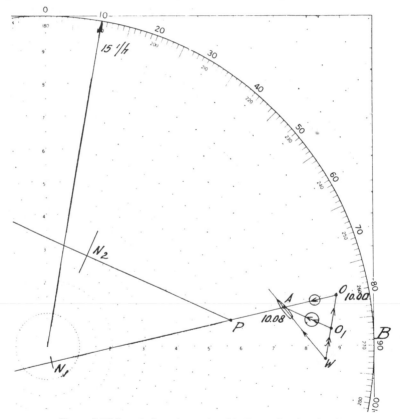

Figure 9.58 Speed alteration to let ship B pass 3 miles ahead

It is difficult to construct the situation during the period of reduction
of speed, but a simple approximation can be made as follows. We assume
that during the first half of the period in which we reduce speed own ship
still has the old speed, and during the other half she has the new speed.
(A similar method may be applied to course alterations.) In this case we

may therefore assume that at 1016 own ship suddenly reduces speed. Until this time we travel the line AP. This is eight minutes and, in this case, AP is equal to OA. We then note point P and call this the *effective point*. Here begins the new line of approach according to which ship B must pass 3 miles ahead of us. From P we draw a line tangential to the three-mile circle.

If ship B keeps the same course and speed, the line WA in triangle OAW will not change. When we now draw the new relative approach line from A parallel to the line PN_2, the line WO is cut at O_1. In the triangle O_1AW the line WO_1 will then indicate the new speed, which proves to be 7 knots.

It is, of course, also possible to choose a new speed and to determine the line of approach and closest point of approach for this speed. The method is similar but the order of performing the operations is reversed.

Course alteration

Next, consider an avoiding action by means of course alteration. In *Figure 9.59* own ship's course is 330° and her speed 10 knots. At 0900 ship B

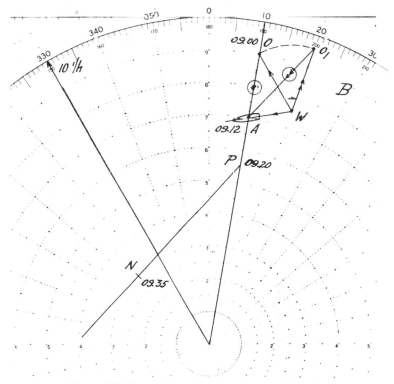

Figure 9.59 Course alteration to pass astern of ship B

was observed at 10° true and 9 miles away. At 0912 with the same bearing its distance is 7 miles. A triangle OAW is constructed as before. It is then decided to pass astern of B by course alteration, the closest point of approach to be 3 miles. We assume that our course is changed at 0919 and that we steer the new course at 0921. Now we must determine what this course should be.

The effective point P, which falls at 0920, is found by advancing the line OA 8 minutes. From P we draw a tangent to the 3-mile circle. This is, therefore, the required line of approach. In the triangle, the line WA remains unchanged; the line WO changes its direction, but its length remains the same since our speed does not change. From A a line is drawn parallel to the new line of approach. We now have two sides of the new triangle and the length of the third, viz. WO, is known. The line WO must, of course, be turned about W to cut the new line at O_1. WO_1 is then the new course, which we are going to steer at 0919. The time of point N can be determined with the speed O_1A. This time will be 0935.

Examples of plotting

1. Own course north, speed 12 knots; radar observations:
   1000: true bearing 20°, distance 10′
   1010: true bearing 30°, distance 6.9′
   Find: course and speed of the other ship and the aspect at 1010.
   Answer: See *Figure 9.60*, course south, speed 8½ knots, aspect 30° green
2. Own course 340°, speed 12 knots; radar observations:
   0800: true bearing 290°, distance 10′
   0810: true bearing 288°, distance 7′
   Find: course and speed, aspect at 0810, closest point of approach and time of CPA.
   Answer: See *Figure 9.60*, course 74°, speed 12.6 knots, aspect 34° green; 0.8′ at 0833.
3. Own course 270°, speed 12 knots; radar observations:
   0400: true bearing 220°, distance 10′
   0410: true bearing 221°, distance 7′
   Find: closest point of approach and time of CPA.
   At 0412 the speed was reduced to let the other ship pass 3′ ahead. We estimate that own ship has this reduced speed at 0418.
   How much is this speed and at what time is this CPA of 3′?
   Answer: See *Figure 9.60*. At 0434 distance 0.5′; new speed 3.5 knots, at 0435 distance 3′
4. Own course 140°, speed 12 knots; radar observations:
   1850: true bearing 190°, distance 9′
   1900: true bearing 190°, distance 7′

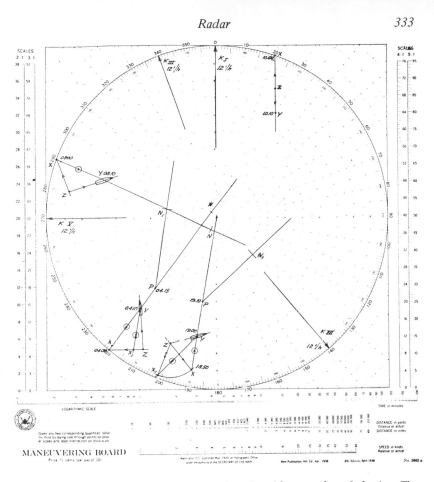

Figure 9.60 Plotting sheet (manoeuvring board), with examples of plotting. The letters XYZ are used instead of OAW

Find: closest point of approach and time of CPA.

At 1909 own course is altered to pass 3' astern the other vessel and at 1911 own ship is assumed to have the new course. Own speed remains the same. What is this new course and at what time is the distance 3'?

Answer: See *Figure 9.60*. Collision at 1935. The new course should be 204° and at 1922 the CPA is 3'.

Reflection plotter

Most radar display units are now fitted with a detachable optical system, mounted in front of the screen, on which the position of other ships may be taken directly from the picture without parallax. Plotting can be effected directly with this optical system. The system is called a reflection plotter and, as shown in *Figure 9.61*, consists of a curved disc S_1 and a flat plate S_2. S_2 passes about 90 per cent of the yellow light of echo P_1 on the screen S_3 of the c.r.t., so the observer at X_1 can see it clearly. If, with a special red pencil, a dot P is made on S_1 exactly above P_1 then the same plate S_2 will reflect this red dot via PRX_1; to the observer, therefore, P_1 and its image coincide. This is also the case if the observer looks from any other direction, e.g. from X_2. Thus in plotting on S_1 parallax is avoided. It is, however, necessary that S_2 should be in the right position relative to S_1 and S_3, and that S_1 and screen S_3 have the same curvature.

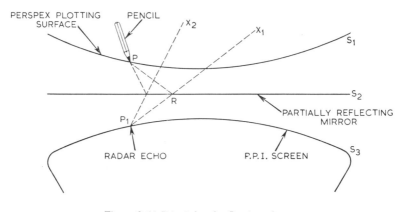

Figure 9.61 Principle of reflection plotter

S_1 is illuminated from the outer edge. By adjusting the intensity of this light the red dots on S_1, which are seemingly on the screen, can be made to disappear, allowing undisturbed observation of the display.

The glass disc S_1 is sometimes rotatable, so that when the ship's head-up picture turns as a result of a change in course, the same plot may be used by rotating S_1. This is done automatically in some types by the gyro-compass.

A reflection plotter is desirable in true motion as well as in relative motion. By plotting several successive echo positions, the direction of movement of the trace of a ship is seen better than by observing the tail. The red dots and lines on S_1 can be wiped off easily with a soft cloth, sometimes soaked in a special solvent.

A disadvantage of the optical system described is that the concave surface of S_1 makes the construction of, say, a plotting triangle rather difficult. Nowadays, however, there are systems, operating on the same principle, where the plate S_1 is flat.

Raytheon radar TM.CPA

Because few personnel are available on the bridge of merchant vessels there is not enough time, in areas with much shipping traffic, for manual plotting in the way described above. There are some makes of radar where plotting is facilitated electronically. (There are also many advanced radars equipped with a computer, which calculates all data required for anti-collision purposes. These radars mostly form part of an integrated anti-collision and navigation system, and will be dealt with therefore in Chapter 10.)

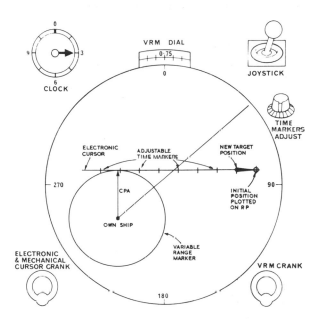

*Figure 9.62 Electronic plotting facility of the Raytheon TM.
CPA radar*

One such make is the Raytheon TM.CPA radar, which has (*Figure 9.62*):
(*a*) a reflection plotter;
(*b*) a variable range marker (VRM), with a scale (above screen) and a crank for rotating the VRM;
(*c*) an electronic cursor (EC), with a joystick and a control to adjust time markers on the EC;
(*d*) a clock (0–12 minutes).
The joystick can be moved in any direction; the horizontal and vertical coordinates of the EC's point of origin change in accordance with the horizontal and vertical components of the joystick's movement.

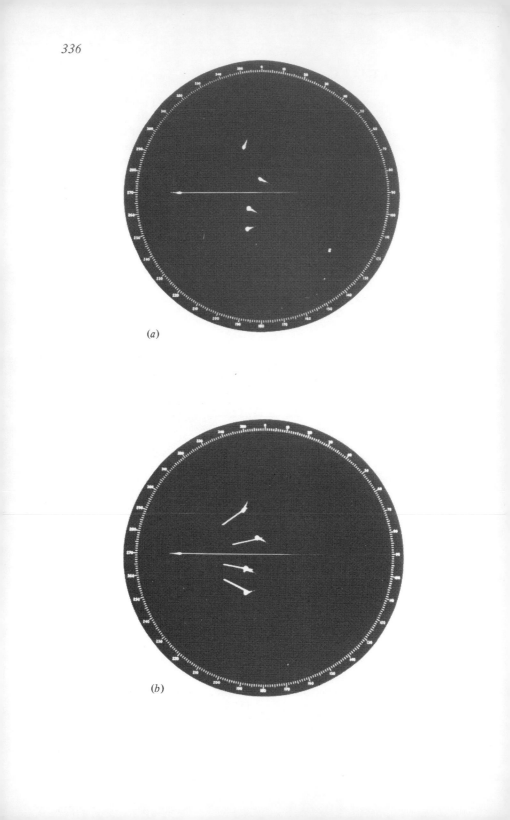

(a)

(b)

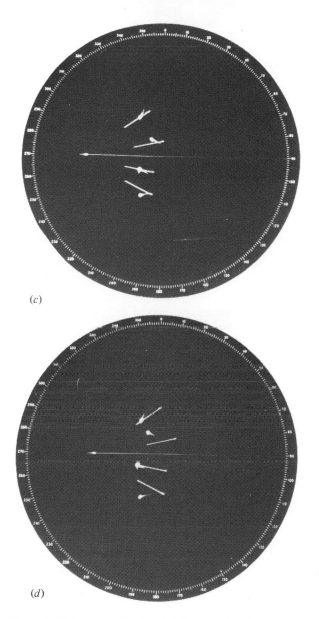

(c)

(d)

Figure 9.63 Decca AC-Radar. (a) True motion shows true
situation. (b) Markers are positioned with bright spots on
potential collision risks. (c) Two of the echoes continue to
move down their markers and are clearly collision risks; the
other two move away from their markers and are not risks.
(d) Two of the echoes will pass well clear of own ship, the
other two have stayed on their markers

The procedure for measurement of the CPA and TCPA on a relative presentation is:

1. Mark target position on reflection plotter. Start plotting clock.
2. A short time later (e.g. 3 minutes) place point of origin of EC on initial plot position with joystick, and rotate EC with cursor crank to intersect target.
3. Run out VRM to touch EC.
4. Adjust marker pips so that the first one falls on the initial plot and the second one on the target in its new position.

The VRM scale reading then gives the CPA distance (say, 0.75 n. miles). The time elapsed between the initial and new target positions, i.e. between time-marker pips, is 3 minutes (the clock reading); if the number of pips between the new position and the CPA is 6.1, the TCPA is 6.1 × 3 = 18.3 minutes.

By other procedures the CPA and TCPA can be measured for a true motion presentation, and the true course and speed of another vessel can be determined.

Anti-collision radar of Decca

The great advantage of a relative presentation is that a check for collision threat can easily be made following any change of bearing of another ship. True motion sacrifices this advantage in order to acquire a better survey of the traffic situation; moreover, it gives an early warning of course alterations by other ships.

The anti-collision radar (AC-Radar) of Decca is a true motion radar but also shows whether the bearing of other ships is changing. Thus this radar combines the advantages of relative and true motion presentation. *Figure 9.20* shows the extra controls for anti-collision radar, above the display.

Information about the bearing is obtained by *electronic markers*. One end of each EM has a bright spot. This spot should be positioned on an echo. The other end is always directed automatically towards own ship.

In *Figure 9.63(a)* four echoes are shown and in *(b)* the markers are positioned. Up to five markers can be placed. If own ship moves over the screen, each EM moves with the ship, parallel to itself, and thus remains directed towards her. If the other ship's bearing does not change, its echo moves down the marker towards own ship; so action has to be taken. This applies for two of the four ships: see *Figure 9.63(c)*, and for a further stage *Figure 9.63(d)*. The remaining two echoes move away from their markers. These ships are not on a collision course and will pass astern or cross ahead. (In general such echoes may already have passed astern or crossed ahead.)

The movement of own ship and other echoes on the screen is in true motion, showing the actual situation; the movement of the echoes along the EM shows collision risk. If an echo moves off its EM there is no risk of collision. If we wish to know the relative motion of the echo we can draw on the reflection plotter a line going through the bright spot of the marker and through the new echo position. This line shows the echo's relative motion. If the radius of the variable range marker is adjusted to touch this line, the point at which it touches is the CPA, and the CPA distance can be read from the VRM scale. In most situations, however, one can see whether there is a danger of collision without drawing this line.

Of course, the AC-Radar cannot be compared with the radars equipped with computers, which are about ten times more expensive, but it is a simple, inexpensive and effective solution for a combined true and relative motion presentation; it makes manual plotting superfluous and does not present too congested a picture.

RADAR REFLECTORS

Detecting buoys, especially in bad weather, is difficult because of the sea-return and, in principle, radar cannot distinguish between the echo from a wave and that from a buoy. Also, because of their shapes, buoys are generally poor reflectors; for example conical ones, especially spar-shaped,

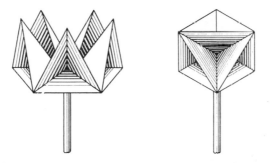

Figure 9.64 Radar reflectors consisting of clusters
of corner reflectors

though cylindrical ones are better. If the reflecting capacity of a buoy can be increased it will stand out better from the sea-return. This may be achieved by fitting the buoy with 'corner reflectors' consisting of three plane surfaces at right-angles to one another. In *Figure 9.47* several examples of the reflections from such arrangements were given. Regardless of the angle of incidence, radiation is always reflected in an exactly opposite direction, to ensure that a great deal of energy gets back to the ship's scanner.

Figure 9.64 shows some types of radar reflectors. The sides of the triangles formed here measure about 400—500 mm. The reflecting power

Figure 9.65 Trinity House buoy fitted with octahedral reflector

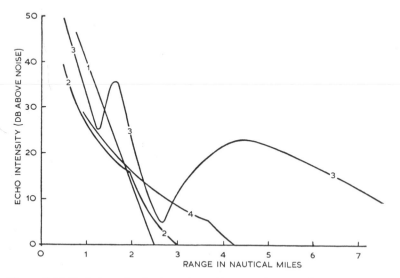

Figure 9.66 Variation of echo intensity with range (courtesy Institute of Navigation)

of such a radar reflector, provided that it is mounted sufficiently high, is greater than that of a ship of about 1000 tons. At present radar reflectors are situated in many places, and are indicated on charts. *Figure 9.65* shows a buoy equipped with reflectors.

Figure 9.66 shows the echo intensity of very strong sea-echoes (curve 1) and of a second-class buoy (curve 2). By mounting a 600 mm radar reflector on the buoy (curve 3), the echo intensity at short distances will become practically equal to that of the strongest sea-echoes while the maximum distance increases from about 3 up to about 9 miles. The reason for this buoy giving weaker echoes at 3 miles than at 5 miles, and weaker ones at 1¼ mile than at 1¾ mile, is reflection from the sea's surface (as shown in *Figure 9.36*).

Waves on the sea surface make reflection more diffuse. The above-mentioned minima at 1¼ mile and at about 3 miles do not appear then. When it is foggy there is often no wind and as a consequence the sea's surface is flat with a specular reflection; as a result, buoys may disappear entirely or partially on approach.

It is very important to know that small or wooden craft (fishing-vessels, yachts, etc.) are observed much more easily if a radar reflector is hoisted on to the mast. A lifeboat equipped with a 400 mm collapsible reflector increases the range at which it is visible on radar from 3 to 7 miles. Light-houses also are sometimes equipped with a cluster of radar reflectors.

RADAR BEACONS

The radar reflectors described above are *passive* devices. Echo energy may also be considerably increased by using *active* means.

A small pulse transmitter is mounted for this purpose on the object, radiating one pulse in all directions as soon as it receives a radar pulse from a ship. The radiated energy of this pulse transmitter will be considerably greater than the echo energy of this object without the transmitter. The great advantage of such pulse transmitters is that they permit identification of certain points, e.g. of the coast.

This active device is called a responder beacon or *racon*, from the first and last syllables of *ra*dar bea*con*. This principle is also called *secondary radar*.

The pulse transmitted by the racon after being triggered by the ship's radar pulse is received on board only if the ship's reflector is pointed towards the beacon. Therefore the beacon becomes visible on the screen at the spot corresponding to its true location. In fact the distance on the screen is somewhat greater since there is a short time delay in the beacon between reception and transmission. The beginning of the beacon signal on the screen of a ship's radar is therefore reproduced at a greater distance than the echo (if the echo is visible); see *Figure 9.67*.

The difference in distance between the echo and the beacon signal is not equal for all beacons; it may be, say, 75 metres. Since we do not know this difference accurately we should use the echo, rather than the pulse behind the echo, for determining the distance to the beacon. The pulse may indeed be used for the bearing, but its primary function is to *identify* the beacon; this is because it is possible for the racon's transmission to be in the form of a pulse-coded signal (e.g. short and long pulses) that will appear on the screen and enable users to identify the racon.

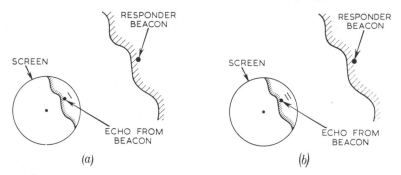

Figure 9.67 Racon echo and identification: (a) racon transmits one pulse; (b) racon transmits two pulses

Volume 2 of *The Admiralty List of Radio Signals* contains a list of the existing radar beacons with their characteristics.

Another type of radar beacon is that which transmits pulses according to a special time schedule, rather than as a reaction to received radar pulses; see *Figure 9.68*. The disadvantage of these beacons, called *ramarks* (from

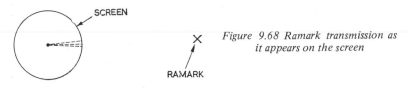

Figure 9.68 Ramark transmission as it appears on the screen

the words *ra*dar and *mark*er), is that they make the picture too congested and so may mask other important objects. Therefore they are now hardly in use, and we shall hereafter deal only with racons.

Racon frequencies

Another difference between racons is the frequency on which they receive and transmit their pulses.

A radar transmitter does not transmit on one frequency but simultaneously generates a number of frequencies, which together form a frequency-band. This band must lie within one of the much broader

Marine Radar bands that have been internationally allocated. The principal Marine Radar band is the 3-cm band (there are very few beacons operating in the 10-cm band).

Racons may transmit their pulses on:

(a) The same frequency as the ship's radar.

(b) Another frequency, preferably at the edge of the Marine Radar band or (if allowed in the future internationally) on one or two special frequency bands. In order to prevent the racon from missing ship's radar pulses, the beacon receiver must have a bandwidth equal to the complete 3-cm or 10-cm Marine Radar band.

(c) A 'swept frequency'; i.e. the transmitted beacon frequency alters gradually, thereby sweeping the whole Marine Radar band in, say, 75 seconds (sweep period). In this case the racon only transmits response pulses when its receiver and its transmitter are momentarily tuned to the frequency of a particular ship's radar and, moreover, the beam of the ship's radar is directed towards the racon. If the Marine Radar band is swept by the racon in 75 seconds, the racon pulses will become visible aboard every 75 seconds during a few revolutions of the scanner.

On account of its low aerial the range of some beacons is rather short, e.g. 5 n. miles. (The explanation for this can be found on page 305.) Hence the sweep period should not be too long, otherwise the navigator would notice the beacon too late. The sweep period, the range of the racon and the speed of the ship may not be considered therefore to be independent of each other.

In methods (a) and (c) the ship's radar need not be altered in order to receive the beacon signals; in method (b), however, the receiver has to be equipped with an extra local oscillator to allow reception on the special racon frequency. A great advantage of method (b) is that the navigator can choose *either* the normal radar picture without racons *or* a picture showing the racon signals only. In the future it will also be possible to obtain the normal picture without racons and, simultaneously, on a special screen, a picture with racons.

When only racons are being displayed the picture is completely free from sea- and land-clutter; this is a great advantage. It can be explained as follows. The clutter originates from reflections of the ship's radar pulses against waves on the sea-surface, land targets, etc. The echoes of these pulses have, of course, the same frequency as the transmitted pulses. The ship's radar receiver, however, cannot receive this frequency if it is tuned to the racon frequency and not to its own radar frequency.

Possible future applications of racons

Due to the stimulation given by IMCO (Inter-Governmental Maritime Consultative Organization) it is probable that racons will play a more important part in the future. Some existing, and possible future, applications are:

(a) to improve ranging on, and identification of, inconspicuous coasts;
(b) to provide identification of coasts that permit good ranging but are featureless, or coasts covered with ice, which alter their contours in wintertime (e.g. in Sweden);
(c) to aid identification of, for instance, a lightvessel in a congested area where many echoes appear on the screen;
(d) as a landfall identification;
(e) to provide a guide to a specific point or into a channel or harbour (the bearing of approach being established from the chart);
(f) to enable a buoy equipped with a depth sounder to indicate depth by means of coded radio pulses;
(g) to indicate bridge piers.

Echo enhancers

Basically there is no difference between a racon and an 'echo enhancer'. After reception of a ship's radar pulse, an echo enhancer transmits one pulse, preferably not longer than the pulse received. On the screen of other ships the enhancer therefore increases the echo caused by simple reflection. In fact this is only the case if the transmitted enhancer pulse coincides with the echo.

Echo enhancers can be used on small and/or low (and therefore badly reflecting) craft, e.g. pleasure boats. Without enhancers these boats would be perceived less well, or only at a short distance, on the screens of other ships. Sometimes their echoes are not reproduced when they are at a longer distance, outside the sea-clutter area, whilst at a shorter distance they might be swamped in the sea-clutter. In such cases an echo enhancer may make it possible for the boat to be detected when it is still outside the clutter area.

Identification of wrecks

A beacon transmitter can be used to indicate the existence of a temporary navigation hazard, e.g. a wreck. This is especially useful when the wreck has not yet been mentioned in 'Notices to Mariners'. It is desirable that in the future a coded signal should be internationally agreed, to be used exclusively for this purpose.

Transponders

Transponder (contraction of *trans*mitter and re*sponder*) is a general term for a radar system that, upon reception of coded or uncoded pulses,

transmits one or more coded or uncoded pulses. Transponders are used for many purposes, e.g. for measuring distances for geodetical purposes. In the future transponders on board ship (including lifeboats) may be used to transfer navigation information by means of coded pulses, for example:
(*a*) identification of the ship;
(*b*) intended course and/or speed alterations;
(*c*) information to assist in search and rescue operations;
(*d*) berthing information;
(*e*) position surveillance of floating seamarks;
(*f*) identification of Ocean Data Acquisition Systems (ODAS).

FALSE ECHOES

False echoes caused by reflection

Echoes on a radar display may sometimes indicate the presence of objects where they do not in fact exist. In *Figure 9.69(a)*, for example, because of the reflection of radar pulses from the funnel, the target may be struck when the reflector is aimed not at it but at the funnel. Part of the energy reflected by the object could then get back to the scanner by retracing its path. Thus the object becomes visible on the picture in the direction corresponding to the instantaneous position of the reflector, and so a false echo is caused in that direction. A similar false echo can occur through reflection from the foremast (*Figure 9.69(b)*).

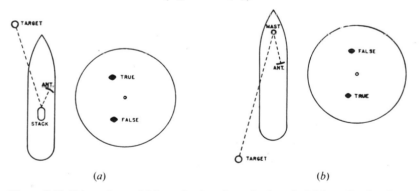

(*a*)   (*b*)

Figure 9.69 False echoes: (a) by reflection from the funnel; (b) by reflection from the mast

Other nearby objects that reflect strongly, such as buildings or tanks along a river, can give false echoes in this way. Their appearance is thus not due to any fault in the radar installation. As a rule these false echoes are in the direction of the blind sector, which is an aid to their identification as such. If the short distance between reflector and funnel is ignored, the range of these objects is accurately represented.

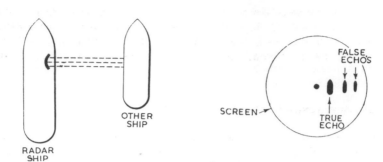

Figure 9.70 False echoes: multiple reflection between own ship and a nearby ship

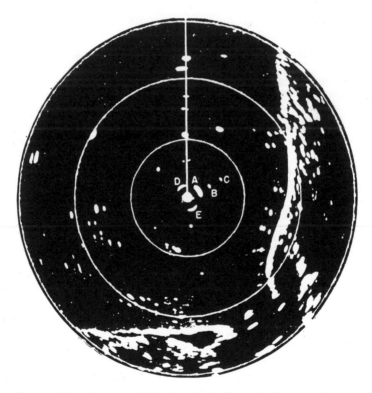

Figure 9.71 Display of Thimble Shoal Channel, Hampton Roads, Virginia. False echoes B and C are caused by multiple reflection from ship A; false echoes D and E are caused by side lobes (courtesy Sperry Gyroscope Co. Ltd)

It is possible to eliminate these false echoes by providing the funnel or other object with a sloping metal plate that reflects the radiation upwards so that no echoes can be received. A similar effect is obtained with the aid of corrugated sheet iron, which scatters the radiation in all directions.

False echoes caused by multiple reflection

Multiple reflection may be set up between own ship and a nearby ship, as shown in *Figure 9.70*. The echoes will appear at regular distances on the screen, and the second and all subsequent ones are false. Echoes B and C in *Figure 9.71* are caused in this way, due to the radar pulses having been repeatedly reflected from ship A.

Echoes from remote objects

With a pulse-repetition frequency as high as 4000 (as used, for example, in the RCA type CRM-N1A-75), successive pulses are produced at intervals of 1/4000 of a second. During this time the pulse and echo can travel 40.5 nautical miles. The echo from an object more than half this distance away (over 20.25 miles) is therefore received only after the transmission of the following pulse. The electron beam of the cathode-ray tube has moved during this period from the centre to the edge, registering incoming echoes, if any. It is not until its *next* trace from the centre to the edge that the beam registers the echo from the distant object, which therefore appears at a false distance. Such echoes are called *second-trace* echoes. (Sometimes third- and fourth-trace echoes have been observed.)

When the p.r.f. is 4000 there is a difference of 20.25 miles between the actual and the indicated distance. Objects located at distances of 20.5, 21 and 21.75 miles will be represented as being 0.25, 0.75 and 1.5 miles away respectively. An object located 20.5 miles away is thus visible on the ¾, 1½ and 3 nautical mile ranges at a distance of ¼ mile. On the longer ranges the phenomenon does not occur with this type since the p.r.f. for these ranges is less, e.g. 1000.

In equipment having a slightly different p.r.f., a simple calculation shows at what distances the objects must be located. Because the p.r.f. is usually lower, these distances are greater than in the example. Second-trace echoes may cause faraway objects (e.g. a coastline), in the presence of super-refraction, to appear on the screen with every point represented in the correct direction but at much closer range. As a result *the shape of the coastline on the screen will appear quite different from the chart. Figure 9.72* shows the chart picture and the radar picture in a particular case. The points A, B and C of the coast at distances of 22, 21 and 22.25

miles will be represented with the correct bearing at distances of respectively 1.75, 0.75 and 2 miles. The great degree of distortion is apparent from the figure. Clearly another ship that moves, for instance, from D to E in *Figure 9.72* will follow a curved path *on our screen* (own ship's movement is left out of consideration here).

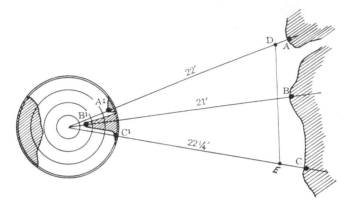

Figure 9.72 Second-trace echoes

In practice, difficulties arising from this type of false echo are less than would be supposed. In the first place, false echoes are much weaker than true echoes in the same place on the screen; they are consequently more vague. Secondly, super-refraction by which echoes are received from objects that are so far away occurs only very occasionally. Thirdly, in most installations the duration of the pulse on the shorter ranges is less, as a result of which the transmitted energy, and also that of the echo, is weaker. Objects located at a very great distance change their bearings only slowly, and so do their echoes on the screen; this enables them also to be identified as false.

Side lobes and false echoes

The side lobes of radiation from the scanner (previously mentioned) can also bring about false echoes (*Figure 9.73(a)*). Under certain conditions it can happen that, when in reflector position OA, ship S is struck by one of the side lobes. The echo produced may be so powerful that it is reproduced on the screen. At this moment the path of the electron beam (timebase) is OA_1 (*Figure 9.73(b)*), and so S becomes visible in the direction of OA_1. A little later when the scanner is in position OB (*Figure 9.73(a)*), the other side lobe strikes ship S and a second false echo is produced in the direction OB_1. In this way *two* false echoes can be obtained, symmetrical to the true echo in the direction OC_1. As a rule there are more than two side lobes and the number of false echoes is then increased. Together they can form a partially interrupted arc of a circle. In practice, however, the energy of the side lobes is so low compared with that of the main beam that these false echoes appear only sporadically, and only from

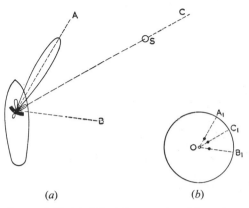

(a)   (b)

Figure 9.73 (a) Side lobes causing false echoes;
(b) the picture on the screen

objects that are nearby and strongly reflective. They can sometimes be made to disappear by manipulating the anti-clutter and gain controls.

The false echoes D and E in *Figure 9.71* are produced as a result of reflection by ship A in the way discussed above.

Radar jamming

A further case of false echoes is radar jamming by another ship. *Figure 9.74* shows a typical example of this kind of interference. It can be experienced when there are other ships using radar in the vicinity; usually, however, this situation does not result in interference. In the first place, the frequencies of the radar sets in the 3 cm band usually differ sufficiently to avoid a receiver picking up radar pulses at a different frequency. Secondly, pulses can only be received in a direct way if a main beam or a side lobe of another scanner is directed at a main beam or side lobe of own ship's scanner; if radar jamming does occur, it is generally caused by another radar's pulses being reflected (by the sea-surface, etc.) into own ship's scanner. Radar interference may be expected more in the neighbourhood of land than at sea.

On account of its typical appearance this interference can be easily recognized. The dots and dashes do not always appear in the same place on the screen and may sometimes disappear during one or more revolutions of the scanner.

HARBOUR RADAR

A further application of radar is the system of one or more shore-based radars (sited at harbour approaches, for example) to supply, in cases of insufficient visibility, incoming and outgoing ships with the information that they are unable to obtain themselves, even when they are equipped with radar. This application is called *harbour radar*. Depending upon the purpose, one speaks of 'shore-based radar for shipping' or 'harbour surveillance radar'.

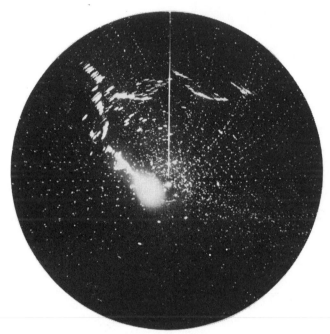

Figure 9.74 Display showing radar jamming

It is not the purpose of harbour radar to guide the ship's navigation from the shore. The full responsibility remains with the ship's master. The harbour radar station only gives information as to the presence of ships at anchor or on an opposite course, etc., and in this way assists safe navigation and cuts down delay (with respect to tides, for example).

Harbour radar enables the port authorities (pilot service, harbour master, etc.) to be informed of the ships in their area and, moreover, provides an immediate means of checking the location of buoys, light-vessels and so on. The position of any collision can be determined at once, and by giving directions to lifeboats and salvage vessels, effective assistance can be supplied in the shortest possible time.

○	off		heading marker alignment
⊙	radar on		range selector
	radar stand-by		short pulse
	aerial rotating		long pulse
	north up presentation		tuning
	ship's head up presentation		gain

Figure 9.75 Symbols for radar switches and controls (a)

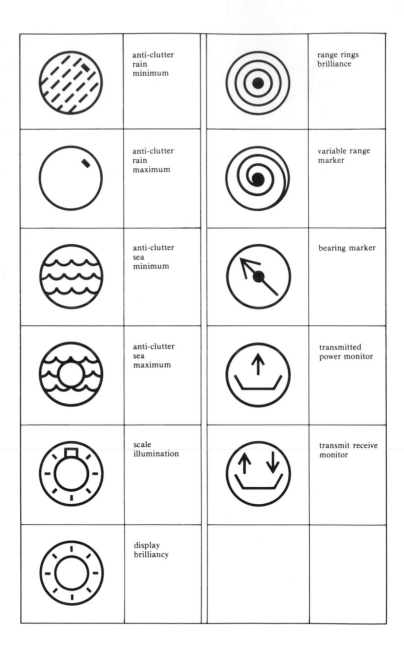

	anti-clutter rain minimum		range rings brilliance
	anti-clutter rain maximum		variable range marker
	anti-clutter sea minimum		bearing marker
	anti-clutter sea maximum		transmitted power monitor
	scale illumination		transmit receive monitor
	display brilliancy		

Figure 9.75 Symbols for radar switches and controls (b)

Besides, there is the possibility of determining under all conditions whether the channel is clear of obstructions, such as ships at anchor, etc. Even if the ship concerned carries radar, it may be impossible to get an overall view of the situation from the ship since factories, dockyards, and cranes may screen and/or cause undesirable reflections to the extent that no clear picture can be obtained from the desired section of channel. However, as harbour radar assistance should be considered an auxiliary aid to navigation, the ship's own radar should stay in operation.

A harbour surveillance radar system may consist of one or more stations. Where several stations are used the ship is passed from one station to the other so as to remain continuously under observation. Since the areas covered by the stations usually overlap, continuity is assured. This is called a block system.

SYMBOLS

The Inter-Governmental Maritime Consultative Organization (IMCO) have accepted a recommendation on the symbols that should appear on radar switches and controls; these are illustrated in *Figure 9.75*. The symbols for radio and radar stations used on British Admiralty charts are shown on page 370.

10

Integrated navigation systems

World losses of ships increased from 157 ships (675 054 tons gross register) in 1968 to 179 ships (1 078 523 tons gr. reg.) in 1973. In view of the increasing density of shipping to be expected, this alarming increase may continue. According to statistics and investigations of the US Coast Guard over several years up to 1970, 73 per cent of collisions are caused by personnel errors in judgement. In 1968 the US National Transport Board conducted a study which indicated that inability to interpret radar observations rapidly enough and mistakes in reading radar data are major causes of collisions. A study by the Chamber of Shipping of the UK postulates that bridge personnel become saturated with data in heavy traffic and bad visibility.

This saturation can be reduced or avoided by an electronic system that:

(a) removes tedious elements of navigation and improves the accuracy of the results, the most important of which is the plotting of radar observations;

(b) selects from the data, supplied by the electronic navigational aids, those which are most important in a certain situation and presents the selected data in a surveyable and easily interpretable way;

(c) calls attention to dangers of collision or grounding by means of warning signals.

The central part of such a system is a computer to which the electronic navigation aids referred to in previous chapters are connected.

On-shore computers are usually placed in rooms where temperature and humidity are kept constant at a suitable value, and where the air is free from dust as much as possible. On board this cannot be achieved; the computer is placed in a well-ventilated room, and is of a design that can withstand the unusual stresses it will encounter on board ship.

The IBM computer unit comprises three modules: a *processor module*, which performs all the calculations and controls all the devices that provide data for, and display the results of, the calculations; a *disk storage module*, which provides large-capacity storage for all programs and other necessary data, and allows the system to initiate different functions by loading main storage from disk storage without manual intervention; and a radar-navigation *interface module*. The *bridge console* is a separate unit that can be located up to 60 metres away.

The electronic system that meets the requirements (*a*), (*b*) and (*c*) mentioned above is called an 'Integrated Anti-Collision and Navigation System', 'Integrated Navigation System' or 'Integrated Bridge System'. Norcontrol, a Norwegian firm, calls it 'Data Bridge'. The radar and other navigation aids described in previous chapters form part of the integrated system, so that each subsystem is of more value than if the subsystem were used separately. Naturally the system does not take decisions concerning a manoeuvre, for example; this remains the exclusive task of the navigator.

According to statistics, 70 per cent of all collisions take place in rivers and waterways leading to the ship's berths. It is regrettable that integrated navigation systems cannot contribute very much towards improving safety in such areas.

A general description of an integrated system follows, adapted mainly from IBM, Sperry and Iotron data.

SENSORS FOR INTEGRATED SYSTEMS

Figure 10.1 is a simplified block diagram of an integrated navigation system. Information from the various navigational aids is fed to the computer via the input interface. The task of this interface is to convert the information of each subsystem into a form that will be accepted and processed by the computer. The output interface serves in the same way to convert the output of the computer to suit the radar display, the Plan Position Indicator and the auto pilot.

The boxes at the bottom of *Figure 10.1* show the system's sensors, for instance the Decca receiver to indicate the position of the ship, and the gyro to indicate its course. The sensors are:

(*a*) a radar;

(*b*) a log measuring the ship's speed relative to the water (e.g. an electromagnetic log), or relative to the sea-bed (a Doppler log);

(*c*) a gyro compass;

(*d*) a satellite receiver;

(*e*) an Omega receiver;

(*f*) a Decca receiver;

(*g*) a Loran receiver;

(*h*) radio bearings and astronomical observations; the position calculated by the navigator with these methods can be supplied by hand via the keyboard to the computer.

The computer is connected (via the input and output interfaces, respectively) to a keyboard and a cathode-ray tube (or *data display*), the latter resembling a television screen. This makes possible communication

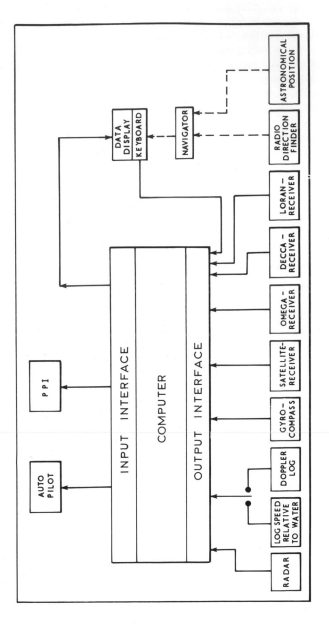

Figure 10.1 Block diagram of integrated anti-collision and navigation system

between man and computer. By means of the keyboard the navigator can give instructions to the system and supply it with data. Via the data display the computer can give the results of its calculations, ask the navigator for data necessary to perform the calculations, and warn him of risks of collision or faults in the system.

The receivers (*e*), (*f*) and (*g*) continuously indicate lane and centilane numbers. The computer converts these into longitude and latitude.

Figure 10.2 Console of integrated navigation system (courtesy IBM)

The ship's radar, with its display unit, is connected to the input of the system, and an anti-collision screen is connected to the output. In some makes, such a screen is in fact a *synthetic radar display*, showing only the echoes and other signals that have been processed by the computer. This screen is called the *Plan Position Indicator* (PPI).

Figure 10.2 shows the console of the IBM integrated system; to the left is the PPI and to the right are the data display and keyboard.

ANTI-COLLISION FUNCTION (IBM SYSTEM)

The principal function of an integrated system is to calculate the course and speed of other ships and to give an early warning of collision danger. For this *collision assessment* (see the flow chart, *Figure 10.3*) the computer requires:

(*a*) the range of the echoes received;
(*b*) the bearings of these echoes, i.e. the direction pointed by the scanner with regard to a reference direction such as the ship's head or true north;
(*c*) the ship's speed, obtained from the speed log or from a manual input via the keyboard;
(*d*) the ship's course, obtained from the gyro compass.

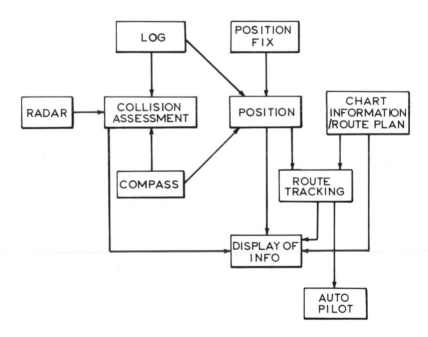

Figure 10.3 Flow chart of IBM bridge system

The computer first checks whether the arriving echoes are persistent or intermittent. Persistent echoes are then checked for size, and those that do not exceed 1800 feet (550 metres) in length are tracked. Tracking consists of viewing the echo regularly in order to check the change of position and calculate course, speed, CPA, TCPA and relative bearing at CPA.

Data about the objects and their movements are shown to the navigator in two ways.

1. On the PPI there appear:
   (a) the normal *radar echo*, obtained directly from the radar receiver;
   (b) a *symbol*, a small triangle above the echo for a moving object and a small square for a stationary object;
   (c) an *identification number* (01 to 42) next to the echo;
   (d) a *vector*. The length of this line, originating at the echo, is proportional to the distance covered within a set time, adjustable from 01 to 99 minutes. The direction of the line is the direction of the echo on the screen. In the 'relative motion' presentation the length and the direction of the vector are therefore relative, and in the 'true motion' presentation both are true. In the latter case a vector can indicate, moreover, the movement of own ship. If another ship performs a manoeuvre its symbol, identification number and vector will flash on and off.
2. On the data display (*Figure 10.4*) certain data appear in tabular form. In this table, the *threat list*, the ships are arranged in order of collision threat, starting with the most serious. The computer tracks all ships

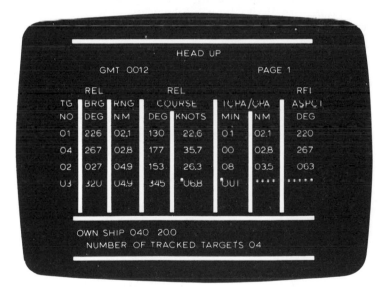

Figure 10.4 Typical threat-evaluation display (IBM system)

within a certain range, but if there are more than 21 ships being tracked it selects for tabular display only the 21 ships that are most dangerous. The data that appear in the table (*Figure 10.4*) are: target number, bearing, range, course, speed, TCPA (time of closest point of approach), CPA, and relative aspect at CPA. Bearing, course and speed will be relative or true depending on the PPI presentation selected. In the table there is room for only six lines (in *Figure 10.4* only four lines). As up to 21 ships may have to be shown, four 'pages' of the display are necessary (three pages of six ships and one page of three ships). Each of these pages can be obtained by pressing the key NEXT PAGE. Normally, however, page 1 is shown because the six ships with the most dangerous approach appear on this page.

Collision threat evaluation and alarm

In the determination of the extent of collision danger, the 'threat evaluation', there are five classes or priorities. Priority 1 is the most serious collision threat. To this priority belong ships within a certain adjustable distance. Priority 5, the least serious, are ships that are moving away from own ship (i.e. ships with a negative TCPA) and that are also outside the range of priority 1. Priority 4 are ships with a CPA greater than a user-defined limit, and priorities 2 and 3 are ships with a CPA less than this limit. The boundary between priorities 2 and 3 is a user-defined TCPA limit. In *Figure 10.5* the CPA set by the navigator is 2 n. miles and the TCPA is 10 minutes.

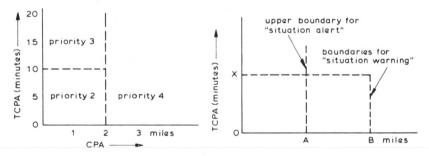

Figure 10.5 Threat-priority relationship   *Figure 10.6 System alarms relationship*
(IBM system)   *(IBM system)*

The number of personnel on the bridge does not allow observation of the PPI and data display to take place continuously. It will often be necessary for the navigator to observe the navigation situation directly visually, or for him to be occupied at the chart table. It is possible that during this time another ship, with a course that was not dangerous initially, has changed her course and brought the CPA within the user-defined limit. In such cases the system with its constant vigilance gives audible and/or visible warning of the collision danger. The cause of alarm is always shown on the data display.

The alarm is related to the priority of the threat; see *Figure 10.6*. There will be a 'Situation Alert' alarm if a ship is within a certain distance A, or a

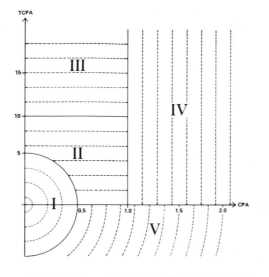

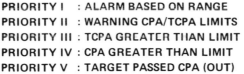

PRIORITY I : ALARM BASED ON RANGE
PRIORITY II : WARNING CPA/TCPA LIMITS
PRIORITY III : TCPA GREATER THAN LIMIT
PRIORITY IV : CPA GREATER THAN LIMIT
PRIORITY V : TARGET PASSED CPA (OUT)

Figure 10.7 'Situation Alert' and 'Situation Warning' alarm limits

'Situation Warning' if a ship is within a certain distance B and also has a TCPA that is less than an adjustable number of minutes, X in *Figure 10.6*. *Figure 10.7* shows the adjustable limits for 'Situation Warning' and 'Situation Alert'.

SPERRY INTEGRATED SYSTEM

The Sperry system will automatically track up to twenty radar-reflecting objects simultaneously (or, as an option, even more).

Selection of ships to be tracked is made manually. With a 'joystick' control the navigator positions an 'acquisition symbol', a small circle, over the echo of a ship to be tracked and presses the 'Ship Acquire' switch; the system will then accept radar data for that ship for automatic tracking. As confirmation a unique identification symbol immediately appears to indicate the present position of the selected echo; see in *Figure 10.8* the 'lock-in' (or 'eyebrows') symbol. Symbols like the circle and eyebrows are made by the interscan method (see page 289).

After eight radar scans (24 seconds), a dotted line (vector) appears, originating at the selected echo. Its direction is the true course of the other ship and its length is the distance run in 6 minutes. After a further 30 radar scans a *Projected Track Line* is displayed from the ship's present position to the point where a collision with own ship could take place (but only, of course, if our course and speed and the other ship's course and speed did not change). This point is called the *Point of Possible Collision* or PPC.

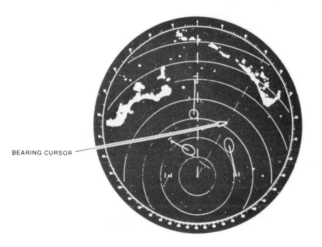

BEARING CURSOR

Figure 10.8 Sperry system display showing acquisition ('lock-in') symbols (two short lines or 'eyebrows'), PPCs, PADs, lines from echoes to PPCs, and (dotted) the bearing cursor. Heading marker intersects a PAD; bearing cursor indicates a safe course (courtesy Sperry Marine Systems)

The PPC is calculated by the computer of the Sperry system on the basis of alterations of radar propagation (or travel) times, etc. In view of the extremely small differences in the sequential measurements, the result of the computer calculation is not sufficiently accurate to determine the PPC exactly. Moreover, it is desirable to observe a margin of safety, in the form of a certain minimum CPA. Therefore an elliptical symbol appears around the PPC. The area inside this ellipse is the *Potential Area of Danger* or PAD; see *Figure 10.8*. The PAD is a danger zone, which must be avoided by own ship. It is determined by the inaccuracy inherent in the system and by the minimum CPA set by the navigator. (If the minimum CPA is set at zero n. miles it remains, however, 300 yards for security's sake.) In order to avoid collisions the navigator has to steer a course that avoids any PAD. An electronic bearing cursor may be used for this purpose.

The Sperry facility described, especially the displayed PAD, makes it very easy to find safe course alterations. This is, of course, true only if the PADs of all ships within range are shown. The system also predicts future PPCs that result from a proposed speed change of own ship.

NAVIGATION OPTIONS (IBM SYSTEM)

Besides the 'collision assessment' facility (pages 358–361) the IBM system supplies other programs to operate the system. The programs are stored in the disk storage unit for ready accessibility, and have the following applications: *position fixing, route planning, route tracking and adaptive autopilot.*

Position fixing

The position mode includes position determination and dead reckoning. For position determination the system accepts inputs from: the Decca system, the Omega system, the Satellite system, the Loran system, and manual entry by the navigator (for instance a position determined by celestial sightings or radio bearings). The system can also caluate the position by measuring range and bearing from a landmark (see below).

Dead reckoning provides a continuous position estimation based on the speed and the course of the ship, the time elapsed since last fix, and the effects of current, waves and wind. At each fix the estimated position is compared with the calculated position. The distance between the two positions is used to calculate set and drift. This information is then used for the next dead reckoning calculation.

Landmark system for position fixing

When approaching a fixed point that is shown on the chart and that can be detected by the radar, the system can calculate the position of the ship with regard to this fixed point. The latitude and longitude of the fixed point must be supplied previously to the computer by means of the keyboard. These points are referred to as 'landmarks'.

When, according to the dead reckoning, the landmark should appear on the screen, the system places a symbol A, B, C, D, E or F on the PPI at the location where the program thinks the landmark should be in relation to the ship. If the letter is not superimposed on the landmark, the navigator can move it with the control stick. The system then measures bearing and distance and calculates the ship's position.

Route planning

The purpose of the route planning program is to give the user the facility
to store his navigation plans in such a manner that he can retrieve them
immediately for his or the computer's use. The system can store up to
99 *routes*, each having eight *legs* of any length. A voyage will be represented by one or more routes.

Use of route planning consists of making a series of keyboard entries
to build or modify a route on the data display and store it on the system
disk. For each leg the user specifies:
(*a*) the geographic coordinates of the end turning point;
(*b*) the type of navigation to be used (loxodromic, orthodromic or
composite);
(*c*) the width either side of the track of a lane within which navigation
is judged by the user, as a result of normal chartwork, to be safe.
Courses and distances between turning points (or way points) are calculated automatically.

Route tracking

A ship will, of course, never follow exactly the calculated route. The
system has at its disposal the actual position (from the position-fixing
program) and the route plan; it can therefore calculate the distance of the
ship from the desired track. If this distance exceeds a given number of
miles (to be supplied by the navigator), the system warns him. The
distances may differ for starboard and port. The lane-boundaries, parallel
to the planned route, are indicated on the PPI in order to show the navigator where the ship is in relation to the route plan. See *Figure 10.9*. When
approaching a new turning point, warning signals are given at distances of,
for example, 5 miles and 1 mile from this point.

In connection with the influence of wind, current and waves, course
corrections are necessary in order to keep the ship on the route. These
corrections are calculated by the computer and displayed on the PPI.

Restricted water navigation

In shoal waters the lane width can be specified to exclude groundings. If
the ship crosses the lane boundary the navigator is warned by audible and
visual signals. In the Sperry and IBM integrated systems the boundaries of
the safe area of navigation can be shown on the PPI. See *Figure 10.10*.

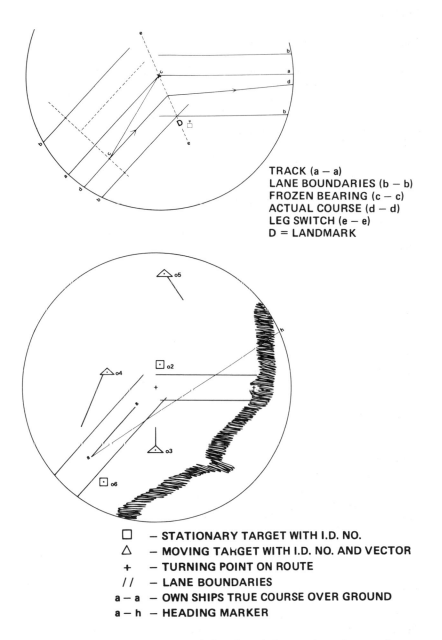

TRACK (a – a)
LANE BOUNDARIES (b – b)
FROZEN BEARING (c – c)
ACTUAL COURSE (d – d)
LEG SWITCH (e – e)
D = LANDMARK

☐ – STATIONARY TARGET WITH I.D. NO.
△ – MOVING TARGET WITH I.D. NO. AND VECTOR
+ – TURNING POINT ON ROUTE
// – LANE BOUNDARIES
a – a – OWN SHIPS TRUE COURSE OVER GROUND
a – h – HEADING MARKER

Figure 10.9 PPI showing ship's track, lane boundaries, turning points, etc. (courtesy IBM)

Adaptive autopilot

The adaptive autopilot controls the steering of the ship. By comparing the gyro compass to the desired course the system can correct the steering engine in order to obtain the required course. At the same time the

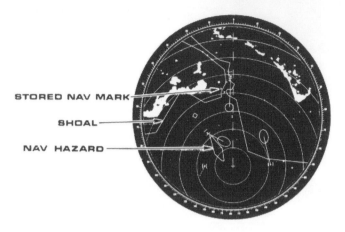

Figure 10.10 Sperry system display showing boundaries of safe areas (courtesy Sperry Marine Systems)

number and extent of the corrections are continuously monitored to determine the best rudder usage.

Use of computer for other purposes

All anti-collision and navigation functions need occupy no more than 25 per cent of the computer's time. It is obvious that the computer may be useful in other activities on board, such as:
(a) engine monitoring;
(b) trend analysis and condition monitoring;
(c) cargo handling;
(d) calculation of wages and other administrative activities;
(e) weather routing.

OPERATION

Figure 10.11 shows the keyboard of the IBM system, which can also be seen in *Figure 10.2*. There are keys for the numerals and for other purposes. The numeral-keys enable, for instance, a position or other data to

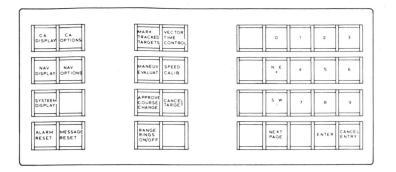

Figure 10.11 Keyboard of IBM system

be entered into the system; if a choice between, say, two possibilities has to be made, this is done by pressing key 1 or 2. The function of the other keys can be seen in an abbreviated form in *Figure 10.11*.

Communication between user and system

In order to make it easy for the user, the IBM system indicates on the data display (*a*) each next step in the operating procedure for a system function, (*b*) warnings, e.g. against collisions, and (*c*) messages of any other kind.

The top line and the two bottom lines of the PPI are reserved for these messages. The top line is for alarm messages and 'system error information messages'. The bottom line is reserved for 'prompt messages', i.e. those that tell the user the next step in the procedure and thus lead him through the desired task. The next-to-bottom line is for information messages and messages confirming the validity of the user's response to a prompt message.

Four examples (of the 50 possible ones) of messages on the top line are:

ALARM THREATENING TARGET, followed by the identification number (01–42) of the ship involved.

NAV MSG XX INVALID SYSTEM POSITION FIX, where NAV MSG means 'navigation message' and XX is a number between 50 and 82, indicating why the determination of the fix is not correct or does not function. For instance, the number 54 indicates that the LOPs (lines of position) are parallel, i.e. have no point of intersection; the number 72 indicates that the distance to a master station of a Decca chain is more than 240 n. miles.

AUTOPILOT HARDWARE ALARM, indicating that the system has detected an error in the autopilot hardware, possibly caused by loss of power or by input/output error.

SHIP IS APPROACHING LANE BOUNDARY.

Examples of a prompt message on the bottom line are:

ENTER DECCA CHAIN NUMBER XX, where XX = 01–39.

ENTER MANUAL POSITION XX.XX.XD/XXX.XX.XD, where the Xs are digits of latitude and longitude (to the nearest tenth of a minute), and D = N, S, E or W.

ENTER MAXIMUM RUDDER COMMAND XX, where XX = 05–35 degrees.

ENTER VECTOR LENGTH IN MINUTES XX, where XX = 01–99.

Examples of messages on the next-to-last line are:

GREAT CIRCLE CALCULATION TERMINATED.

PAGE X REQUESTED, indicating that the user has been asked for the next page of the list of ships with anti-collision data.

MANEUVER EVALUATION ACTIVATED, indicating that the computer is calculating CPA and TCPA for a Trial Maneuver that the navigator has submitted to the computer.

Procedure for routing calculation

By way of example, here is the procedure for routing calculation:
1. Press key NAV OPTIONS (the 'navigation options' are the system functions other than 'collision assessment' or CA). There then appears on the display a list of the four navigation options: 1 = *position fixing*, 2 = *adaptive autopilot*, 3 = *calculate* and 4 = *routing*. The prompt message indicates: SELECT NAV OPTIONS.
2. For routing we therefore have to press key 4. Immediately the 'Route Display' page appears (*Figure 10.12*). To enable calculation of the route we have to supply the computer with some data, filling in the table on the data display. First of all the route should be given a number of two digits for identification in the system.
3. After having pressed the keys 1 and 2, for instance, the number 12 appears on the display, with a prompt message: ENTER START POSITION XX.XX.XD/XXX.XX.XD.
4. The position must be supplied and then the key ENTER pressed. The

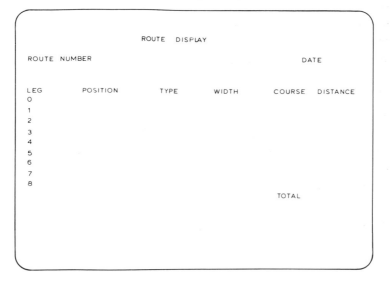

Figure 10.12 Route display of IBM system

position is shown on the display in the second column beside 'LEG 0'. The prompt message is: ENTER LEG (L/L, TYPE, WIDTH) XX.XX. XD/XXX.XX.XD-TT-PP, P-SS, S. Here L/L is the latitude and longitude of the first turning point (or the end of the route if it consists of only one leg). The two digits for the route type TT are 00 for loxodromic (rhumbline) or 90 for great circle. It may be that the navigator wants to avoid areas with too high latitudes, e.g. more than 60°; keys 6 and 0 should then be pressed, in which case the maximum latitude for a great circle route will be 60° PP,P and SS,S are the safe distances in miles to port and starboard (the lane width). If these distances are exceeded the navigator will be given an alarm.

5. Press key ENTER. The data given in step 4 for the first leg are now entered into the system. On the display the first leg definition appears on the display as 'LEG 1'. The prompt message is: ENTER LEG (L/L, TYPE, WIDTH) XX.XX.XD/XXX.XX.XD-TT-PP,P-SS,S. In the same way as before the data for the second leg (and thereafter further legs, if any) should then be entered.

6. After the key ENTER has been pressed for the last leg, courses and distances calculated by the computer will be shown in the two last columns of the table, as well as the total route distance. The prompt message is: SELECT MODIFY = 0, CANCEL = 1, SAVE = 2, MAKE ACTIVE = 3 X. Depending on which of the above options is desired, the key 0, 1, 2 or 3 should be pressed.

Symbols for radio and radar stations, etc. (as used on British Admiralty charts)

RC	*Non-directional radio beacon*
RD RD 269°30'	*Directional radio beacon*
RW	*Rotating-pattern radio beacon*
RG	*Radio direction-finding station*
Radio Mast Radio Tr Radar Tr Radar Sc	*Radio mast or tower; radar tower or scanner (landmarks for visual fixing only)*
TV Mast TV Tr	*Television mast or tower*
R	*Coast radio station providing QTG service*
Ra	*Coast radar station*
Racon	*Radar responder beacon*
	Radar reflector (not charted on IALA system 'A' marks; see L 70)
Ra (conspic)	*Radar-conspicuous object*
Ramark	*Radar beacon*
Aero RC	*Aeronautical radio beacon*
3	*Radio calling-in point, way or reporting point (with number if any) showing direction(s) of vessel movement*
Consol Bn	*Consol beacon*

Index